Infinity, Causality and Determinism

Contributions to Philosophical Theology

Edited by Gijsbert van den Brink, Vincent Brümmer and Marcel Sarot

Advisory Board:
David Brown
Paul Helm
Eberhard Herrmann
Werner Jeanrond
D. Z. Phillips
Christoph Schwöbel
Santiago Sia
Alan Torrance
Nicholas Wolterstorff

Vol. 6

Frankfurt am Main · Berlin · Bern · Bruxelles · New York · Oxford · Wien

Eeva Martikainen
(ed.)

Infinity, Causality and Determinism

Cosmological Enterprises
and their Preconditions

PETER LANG
Europäischer Verlag der Wissenschaften

Die Deutsche Bibliothek - CIP-Einheitsaufnahme

Infinity, causality and determinism : cosmological enterprises and their preconditions / Eeva Martikainen (ed.). - Frankfurt am Main ; Berlin ; Bern ; Bruxelles ; New York ; Oxford ; Wien : Lang, 2002
   (Contributions to philosophical theology ; Vol. 6)
   ISBN 3-631-39146-3

ISSN 1433-643X
ISBN 3-631-39146-3
US-ISBN 0-8204-5490-7

© Peter Lang GmbH
Europäischer Verlag der Wissenschaften
Frankfurt am Main 2002
All rights reserved.

All parts of this publication are protected by copyright. Any utilisation outside the strict limits of the copyright law, without the permission of the publisher, is forbidden and liable to prosecution. This applies in particular to reproductions, translations, microfilming, and storage and processing in electronic retrieval systems.

Printed in Germany 1 2 3 4 5 6 7

www.peterlang.de

Table of Contents

1. Introduction
   Physics, Philosophy, and Theology Facing Cosmological Challenges   7
   Eeva Martikainen

Historical and Metaphysical Concepts of Causality

2. Causality and Cosmology   19
   The Arabic Debate
   Taneli Kukkonen

3. God, Causality, and Nature   45
   Some Problems of Causality in Medieval Theology
   Reijo Työrinoja

4. On the Relationship between Metaphysics and Physics   61
   Olli Koistinen

5. The Concept of Infinity in Kant and Hegel   73
   Jussi Kotkavirta

Quantum Mechanics and Causality

6. The Copenhagen Interpretation of Quantum Mechanics
   and the Question of Causality   89
   Tarja Kallio-Tamminen

7. Subject and Object in the Study of Nature   101
   Sami Pihlström

8. Causes, Causes, Causes   111
   Three Aspects of the Idea of Cause
   Jaakko Hintikka

9. Quantum Mechanics and Emergence — 119
I. A. Kieseppä

Discussing the Theory of Everything

10. Time and Causality in the Theory of Everything — 133
Kari Enqvist

11. The Concept of Space in Relativity Theory and Cosmology — 143
Raimo Lehti

12. Spinoza in Context — 165
A Holistic Approach in Modern Terms
Rainer E. Zimmermann

13. Spinoza and Modern Physics — 187
Comments on Professor Zimmermann's 'Spinoza in Context'
Juhani Pietarinen

Cosmology and Theology

14. Cosmology in Science and Theology — 193
Some Perennial Differences
Stanley L. Jaki

15. An Axiomatic Foundation of Quantum Mechanics — 205
Paradoxes, Causality, and the Unification of Sciences
Juleon M. Schins

16. Cosmology, Theology, and the Question of Ultimate Explanation — 227
Heikki Kirjavainen

Notes on Contributors — 247

Index — 251

1. Introduction

Physics, Philosophy, and Theology Facing Cosmological Challenges

Eeva Martikainen (Helsinki)

1. *Challenges Raised by the New Physics*

Cosmological theories have been unpopular within philosophy and theology for a long time. Postmodern thinking may be characterized by the condemnation of metaphysics, by cultural fragmentation, and by the segmentation of academic disciplines. In particular, positivism, logical empiricism, and post-wittgensteinian philosophy of language have produced not only philosophical relativism but also a skepticism toward the possibility and usefulness of interdisciplinary rationality. Many theologians have also shared these philosophical notions. They have argued that theological language is its own entity and, thus, should be studied only according to its own principles. It has been argued that the importance of religious language is limited to theologians and religious communities. It need not be understood by 'outsiders.' This segmentation of the realms of life and of academic disciplines has had its influence on our contemporary understanding of cosmology as well.

Meanwhile, particularly during the last three decades, speculative cosmology has been transformed into a discipline that follows the rigorous and exact methods of physics. The new popularity of cosmology has been reflected in an expansive growth of related literature. For instance, in the United States the academic and public discussion has been stimulated by Steven Weinberg's book 'Dreams of a Final Theory'. Steven Weinberg received The Nobel Prize in physics (1979) for his efforts to unify weak and electromagnetic interaction between elementary particles. His research has been an enormous contribution to physicists' on-going efforts to unify the four natural forces, namely, strong and weak interactions, gravity, and electromagnetic interaction.

Contemporary physics has been deeply influenced by two conflicting interpretations, namely, quantum physics and the general theory of relativity. The most notable developer of quantum physics has been Niels Bohr, while Albert Einstein shaped the general theory of relativity. A theory that would bring quantum physics and the theory of relativity together is yet to be discovered, even after decades of active efforts. If physicists were successful in their efforts to find common ground for quantum physics and the theory of relativity in the Theory of Everything, would philosophy and theology be able to join the discussion? Would these disciplines have any role in explaining the ultimate nature of reality? This discussion is already challenged by physicists. Some physicists, especially those advocating what is called the Theory of Everything, seem to hold that certain perennial problems in metaphysics can be solved in a fully developed physical theory. The early discoveries of quantum physics in the last century, however, have been paralleled with efforts to develop a specific philosophy of quantum physics. The discoveries of quantum physics have, in fact, shed new light on our understanding of the foundations of reality and they have challenged the mechanistic notions of Western thinking. For instance, unlike the exact values of classical physics, the objects and properties of quantum physics are neither exact nor objective. They interact rather with the very act of measuring; thus such measures as velocity and position of an elementary particle are relative. This interactiveness raises the question whether it is still possible to speak of an uninvolved, objective, observer and whether the role of the scientific subject has totally changed. The classic debates between realism and idealism, and between realism and nominalism, are in this new situation, once again brought to live. The relationship between physics and metaphysics is also not just an antiquarian problem but very actual indeed.

The difficulty for contemporary physics and philosophy to take metaphysics seriously in their cosmological considerations may be because the role of metaphysics in the history of philosophy has been ambiguous. Metaphysics may be either the general investigation into the nature of being itself, or it may be the investigation into the nature of God. Aristotle termed 'the first philosophy' as a science, but this science was later dubbed 'metaphysics.' The object of the first philosophy was being qua being. Aristotle says that the first philosophy should be viewed as the science in which the nature and existence of the unchanging separate substance, i.e., God, is considered. Even if it were true that God is the explanation and cause of all changes and changing things, it does not follow that knowledge of God will include knowledge of objects and changes or that theology will itself be concerned to study the attributes of being qua being. Later development in the history of metaphysics shows in any case that metaphysics is to be the most abstract and general of all sciences: a science which investigates something that is somehow associated with everything there is – viz. being.

These were the questions underlying the colloquium 'Infinity, Causality, and Determinism: Cosmological Enterprises and Their Preconditions,' an international and interdisciplinary colloquium[1], which offered philosophical, theological, and physical approaches to historical and contemporary cosmology. This colloquium

---

1 Helsinki, Finland, 8–9 May 2000.

was organized as a part of 'Theology and Natural Sciences,' a research project that is financed by the Academy of Finland. The colloquium was co-sponsored by CTNS Religion Course Program in Europe and the Department of Systematic Theology in the University of Helsinki. The colloquium 'Infinity, Causality, and Determinism' has examined a long tradition of Western thinking within which has been a consistent effort to study and understand reality as an intelligible whole. The approach was chosen even though many contemporary philosophers and theologians have questioned this traditional foundation of Western thinking.

2. *The Aristotelian Concept of Causality and the Medieval Discussion on Causality*

It is impossible to comprehend fully the contemporary scientific approaches in cosmology without an awareness of the long history of physics and of the philosophy of science. Western scientific thinking has its roots in questions, which were formulated and addressed by the thinkers of antiquity. Similarly, cosmology has a long history to which physicists, philosophers, and theologians have all contributed. I am going to outline a few central events within the history of Western cosmology that reveal interaction, but also controversies, among physical, metaphysical, and theological approaches. Within classical cosmology, metaphysics has followed physics, as can be seen even in the etymology of the term 'meta-physics.' Within this classical framework, philosophical theology has been founded on metaphysics, although theology has also followed another tradition that is peculiar to it, namely, that of Scripture.

In an inquiry concerning the history of the interrelations of physics, metaphysics, and theology, a study of Aristotelian causes is an inevitable starting point, as was seen in the course of the colloquium. In fact, one might say that in the history of Western thinking no single theory of causality has had as much influence as that of Aristotle, as Jaakko Hintikka asserts in his article. Aristotle's conception of causality is inherently connected to his metaphysics and his logic. Aristotle's notion of four causes brings together his cosmology, ontology, epistemology, and logic. Moreover, the Aristotelian notion of the Prime Mover that united the efficient cause (causa prima efficiens) to the final cause (causa finalis) offered a philosophical starting point for theological notions concerning God. Within the subsequent tradition of Aristotelian realism, physics, philosophy, and theology jointly contributed to the understanding of the universe.

The medieval Christian philosophical debates concerning Aristotelian realism and nominalism continued to address fundamental scientific issues of causality and other forms of scientific explanation. As *Reijo Työrinoja* writes, Thomas Aquinas argued that every created being has its own active power by which it can, together with the divine primary causality, act as an immediate cause of physical effects. God is the formal and final cause of every action but not the immediate cause. God does not need to intervene in the actual world at every moment, because substantial forms and secondary causes take charge of their tasks in relatively independent ways. Aquinas was, however, aware of an alternative way of thinking that was in opposition to his own metaphysical and ontological views.

This alternative paradigm came from Arabic philosophy. In al-Ghazâlîs (1058–1111) conception, described by *Taneli Kukkonen,* the most vehement critics of Aristotle's worldview were theologians who saw his holistic conception of reality as deterministic and, thus, problematic from the viewpoint of God's omnipotence and of human free will. In this respect, the early medieval Islamic tradition of Aristotelian interpretation offered two approaches that were to have a long-lasting impact. First, some thinkers argued that God's influence in the world has to be viewed as functioning in a category entirely separate from natural causes. According to this interpretation, Aristotle's causes would be limited, therefore, to the natural world. Second, some thinkers who were inspired by the so-called occasionalist worldview took an atomistic outlook, arguing that God, as the sole and direct causal, agent perpetually creates the world. Islamic interpreters of Aristotle defended Greek philosophy and its search for certain knowledge concerning the transient world by arguing that a holistic scientific description did not necessarily lead to determinism.

3. *Nominalistic Approaches and the Birth of Modern Science*

Medieval nominalists questioned the Aristotelian ideal of realistic and syllogistic science. They were the first to set physical, philosophical, and theological approaches to understanding reality into clearly separate and systematic compartments. They argued that the God who created the world as an expression of His free will must have had several different creative options, thus, the laws of this world do not necessarily lead to a full understanding of the Creator. This idea put the notion of a unified rational reality in to question. If the laws of nature are not identical to logical possibilities, then the properties of created beings are not necessarily limited to their essentialist (Aristotelian) definitions. Accordingly, nominalists argue that 'universals' are but only individuals or mental concepts. Conversely, universal concepts, such as the notions of species and natural laws, are formed only after observation of regular patterns and resemblance in the operations of individual things.

As Reijo Työrinoja points out, William of Ockham argued that no valid scientific demonstration for the existence of causality can be given. We have no intuitive cognition of the causal relation itself or of other relations. Relations are rather inferred by reasoning. For Ockham, however, secondary causes are genuinely explanatory causes in the epistemological sense, although one can have no ultimate ontological guarantee of the causal connection between two phenomena.

The nominalist approach paved the way for a modern science that can be characterized by its empirical study and by its refusal to believe in a priori definitions. In fact, nominalism has had a decisive impact on modern science and its understanding of causality. Lutheran theologians were actually supporting the new science from the beginning. Theological interest concerning the modern science was later given to an astronomer and theologian, Andreas Osiander, who wrote a cautious but in principle approving preface to Copernicus's book. Mathematical astronomy had until the end of the sixteenth century an important place in the University of Wittenberg, influencing other universities in Germany and in Scandinavia. Lutheran theologians were in many ways interested in science and philosophy. Unfortunately, contemporary Protestant theologians have forgotten this heritage of the sixteenth century.

4. *Newton's Mechanics:*
   Questioning the Problem between Physics and Metaphysics

Newton's mechanics, which radically differed from the Aristotelian understanding of motion, marks the breakthrough of modern physics. It may be argued that the concept of cause is more of a philosophical than a physical issue. With the idea that 'force is its cause' (Raimo Lehti) Newton separated the language of physics from the language of philosophy. The natural sciences had recently acquired a status that was independent and conceptually distinct from theology. Central figures in this development were Newton and Descartes. Newton demonstrated that it is possible for us to investigate nature theoretically as a collection of natural laws, that is, as a collection of mathematically formulated functional dependencies (invariances). No reference to teleological or theological factors is needed. Descartes gave this view philosophical credibility. An essential element in Descartes thought was the idea that nature is mathematical at its heart: mathematical invariances are even ontologically prior to the universe itself.

The new independence of physics challenged philosophy and as well as theology. As *Heikki Kirjavainen* describes, the program of natural theology arose in the late seventeenth century as a cognitive problem. This program in turn could not have been launched before the natural sciences gained their independence. The Enlightenment program of natural theology is an attempt to address precisely this new problem: with regard to the mathematical laws of nature, are there any features in nature that enable us to say something about its ultimate foundation?

From here, the notion that both the regularity of nature and some of its special features refer to God was but a short step away. The harmony or beauty of creation could be offered up as such a special feature: 'nature speaks of its Creator.'

Descartes's and Spinoza's use of geometry perhaps followed the best new developments of physics, giving them a philosophical formulation. In Kant's philosophy, science and moral philosophy were given separate a priori foundations, a fact that has had a strong influence on later philosophy and theology. Because the contemporary practice of treating physics, philosophy, and theology as totally separate disciplines goes back to the philosophy of Kant, the issue is studied in several articles (Olli Koistinen, Jussi Kotkavirta, and Sami Pihlström).

The aim of *Olli Koistinen's* article is to consider whether physics and metaphysics should be seen as presenting conflicting claims on the same subject matter, the nature of reality. Even though philosophers do not suggest that their metaphysical considerations can enlarge the scope of pure empirical knowledge, some physicists still imagine a kind of battle between metaphysics and physics. According to them (e.g., Steven Weinberg, a winner of the Nobel Prize in physics), metaphysics is not needed, because physics can be seen to replace metaphysics or to make empty the claims of metaphysics.

Koistinen considers in his article Descartes's and Kant's notion of the relationship between physics and metaphysics. He points out that these philosophers thought that, even though physics as an empirical science can exist without metaphysics, the ultimate justification for the truths of physics comes from metaphysics. Metaphysics may not enlarge the scope of physical truths, but metaphysical justification is needed to transform the cognitio of nature into scientia of nature. Notwithstanding his idea that empirical physics does not need metaphysics, Kant argued that physicists who despise metaphysics despise themselves in not wanting to build science upon a firm foundation.

Many other papers of the colloquium also referred to Kantian idealism. *Jussi Kotkavirta* deals with the different meanings and roles of the concept of infinity in Kant's critical philosophy and in Hegel's absolute idealism. Kant believes that reason and freedom are metaphysically infinite and in his practical philosophy argues that as reasonable and free persons we should adopt an infinite perspective in respect to our actions. Our knowledge, however, cannot reach beyond the mathematical infinity of space and time as the pure forms of intuition, not beyond our categorical apparatus. Hegel wrote that Kant's arguments for the inescapability of the antinomies concerning our knowledge of the infinite fail. Hegel redefines dialectics as a rational procedure of sublating, not establishing, limits and of reaching the absolute totality, i.e. the metaphysical infinity.

Philosophers have to defend the autonomy of the human being (Rainer Zimmermann), and this goal is perhaps not possible without Kant. The idea that the axiomatic foundation of the sciences, moral philosophy, aesthetics, and religion is in the knowing subject is also widely discussed within modern Protestant theology. Whether Kantian philosophy can provide even today a fresh approach to understanding the relationships among physics, philosophy, and theology, is discussed in the papers of *Sami Pihlström* and *Heikki Kirjavainen*. Kant's approach to sciences, however, is implicitly criticized in some articles. For instance, *Juleon M. Schins* suggests an axiomatic foundation of the science of philosophy based on Aristotelian hylomorphism. He is finding this possibility in quantum physics, especially in the universalismus of wave functions.

5. *Quantum Mechanics and Causality*

From the beginning of this century questions concerning causality, determinism, and the rational knowability of reality have been brought under a new kind of focus within physics, especially because of quantum physics. *Tarja Kallio-Tamminen* claims in her article that quantum theory is essentially statistical. The world is not totally predictable, and the classical concept of causality is not available. The conflict between different interpretations among physicists and conceptions of reality culminates in the measurement problem. The traditional dualistic framework of thought has not offered tools for solving the problem. Niels Bohr's framework of complementarity offers one way to avoid the measurement problem. It also removes all apparent paradoxes of quantum mechanics without adding any unnecessary auxiliary hypothesis to the theory. Bohr's approach would, however, entail radical changes to our understanding of reality. Criticizing the subject - object dualism challenges the accustomed foundations of the natural sciences. Descartes proclaimed that neither God nor any rational soul present in the world would ever disturb the ordinary course of nature in any way. Denial of this foundational dualism would probably collapse the paradigm of the modern Western conception of reality. The founding fathers of the Copenhagen interpretation were convinced that quantum theory demanded radical changes to the metaphysical preconditions of traditional physics. Their proposals for profound revisions of our ontological and epistemological approaches to reality are examined against the background of later interpretations that have returned to such basic presuppositions and objective ideals of classical metaphysics as the detached observer, determinism, and reductionism.

The standard interpretation of quantum mechanics contains plenty of paradoxes, as described by Tarja Kallio-Tamminen. In spite of all the research effort invested in this issue during the last decades, the number of inconsistencies seems only to increase. *Juleon M. Schins* argues against the standard interpretation of quantum mechanics that an important omission in the traditional foundation concerns the formulation of the exact relation between wavefunction and a single event. On the basis of Aristotle's hylomorphism he proposes a radically different axiomatic foundation of quantum mechanics, which requires also fewer axioms than the traditional foundation; it solves all the traditional inconsistencies, supposes a non-deterministic worldview with a well-defined notion of causality and is compatible with a theoretical unification of sciences.

Jaakko Hintikka offers in his article a new conception of causality, which is based on quantum physics. He writes that in quantum theory certain pairs of variables, called noncommunitive variables, are mutually dependent. This does not seem to be much of a novelty until one asks: how are dependencies and independencies between variables expressed in a logical language? What is ultimately needed is a more flexible logic. Important issues for the new logic based on quantum physics are the measurement problem and the nonlocality problem. The new logic suggests that mutual dependence of noncommunitive variables is indeed a hallmark of the quantum logic.

But what does the new logic have to do with the notion of cause? If the characteristic dependencies in quantum theory are symmetrical, they cannot be captured by means of the classical asymmetrical concept of cause, which is, therefore, inadequate for the purposes of quantum physics. Thus, it is not surprising that in quantum physics such phenomena occur as nonlocal dependencies that are not causal in the usual sense.

6. *The Theory of Everything:*
   Radical Changes in the Concepts of Causality, Space, and Time?

One of the interesting developments within particle physics has been the emergence of the so-called Theory of Everything (TOE) that would be based on a quantum field theory. So far it has been possible to unify two natural forces, namely, electromagnetic and weak interaction in a quantum field. If general relativity and strong and electroweak interactions are found to be low energy approximations of a unified Theory of Everything, our traditional understanding of causality and time might change radically. For example, if physicists are able to prove the superstring theories that are presently viewed as the most likely candidates for the Theory of Everything, would that force us to reject our traditional notions of time and causality? As *Kari Enqvist* suggests in his article: 'At the most fundamental level, time may, thus, not exist, nor time ordering; hence, in the deepest sense there would be no causality as we define it.' Nonetheless, all these theories that would unify quantum physics and Einstein's Theory of Relativity are yet to be verified. Moreover, even if a Theory of Everything that would unify all four natural forces were eventually mathematically explainable, we might not be able to understand its implications.

The Theory of Everything is discussed in this volume in the context also of astronomy, which has a different approach to the theory of science than physics does. *Raimo Lehti* says, that the physicist looks at space from the perspective of general theories, and the astronomer from the perspective of our unique cosmos. He asks why the differences between these views have not often been clearly stated. Perhaps the reason lies in a reluctance to pinpoint discrepancies in such an excellent theory as the theory of relativity.

If one were to discover a Theory of Everything, what would be its challenge to philosophy and theology? Would a Theory of Everything include ontological and epistemological approaches that have typically shaped philosophical cosmology? If a Theory of Everything allowed a flexible development of space-time, our traditional notions of causality and determinism might radically change.

Today's physical cosmology presents philosophy with challenges similar to these physics posed for classical metaphysics at the turn of the twentieth century. It may be argued that it is possible to move from physics to metaphysics, because all theories of physics include metaphysical presuppositions. In other words, physics must presuppose that an order in the relations among various beings. It is more difficult to invert the direction, however and move from metaphysics to physics, especially if one wants to hold on to rigorously scientific and quantitative correlations of physical theories. One must distinguish between the concepts of physics and those of metaphysics, making it impossible to introduce familiar metaphysical concepts into the realm of physics. Nonetheless, even if physical and philosophical theories use different concepts, the dialogue between physics and philosophy can be mutually beneficial, as argued, for instance by Rainer Zimmermann.

Quantum gravity, as the most promising candidate for a Theory of Everything, as described by *Rainer Zimmermann,* sets out to describe the real world in the philosophical sense: as the foundation of the modal world we are able to perceive. Although utilizing the most recent techniques of mathematics and theoretical physics, quantum gravity reproduces the ancient search for the answer to a very old problem intrinsic in the very structure of human consciousness: 'What is the nature of the discrepancy between the world as perceived (which is constituted by means of mapping the environment) and the world as existing independently of human perception?' In other words: 'What is the precise difference between a picture (or model, or metaphor) of the world and the world as the object of which this picture is made by the process of mapping?' Spinoza was the first philosopher to deal with this problem in a modern language, which is still accessible to us today. Spinoza himself discussed in a nutshell all relevant consequences of his ideas, which pointed to a realist materialism rather than to an idealist philosophy. And it is exactly for this reason that his ideas can be seen as forerunners of those ideas, which are underlying the philosophical aspects of recent research. Zimmermann compares the basic aspects of Spinoza's metaphysics with recent results in the development of a Theory of Everything in modern physics. He shows that Spinoza's outlook is still intrinsic to ongoing research in the sciences.

Twentieth-century thinkers have reconsidered the relationship between philosophy and science, and generally science has been given the priority over philosophy. However, recently a philosophy is being achieved which follows science rather than laying down its grounds. This philosophy is ultima philosophia in that it can still unfold its heuristic potential which supporting scientific research in a movement of critical reflection. Therefore, philosophy can achieve something that the sciences cannot. On one hand, it can establish a contextual correlation between scientific approaches to the fundamental level of the world. On the other hand, it can create an ethical framework for the practice of sciences.

7. *Conclusion: Cosmology and Theology*

What role can theology then play in the contemporary discussion concerning physical cosmology? In the Middle Ages, theologians criticized the Aristotelian holistic worldview, and subsequent theologians were critical about early modern science and its efforts to explain all events in the world on the basis of physical causation. In both eras, theologians were arguing that physical events could also be explained as caused by the omnipotent God. In fact, as Spinoza's philosophy, for instance, has shown, early modern thinkers did not exclude the possibility of seeing God as the prime cause even within a deterministically defined natural system. As the history of theology has shown, theology has been able to contribute fruitfully to cosmological discussions, and I believe that this profitable interaction between theology and cosmology is by no means over. This interaction is not possible without philosophy. Whether the philosophy to be developed will be Aristotelian, Spinozian, Kantian or something very new, we must leave for further discussion. All of these possibilities have been suggested in this volume by means of certain careful argumentation. Important in any case to notice is that no 'theological cosmology as such' can be based on the Bible. What is understood by 'the story of creation' in the Bible will be, and has been, studied in the Biblical context and in the theological sense and not as a scientific explanation (*Stanley L. Jaki*).

One might think that we are left in this volume with very few definite solutions. It may indeed be that the search for answers to the problem of the universe, its origins, and its structure is only at its beginning. The discussion concerning the relationships among physics, philosophy, and theology was certainly one of the most important contributions of the colloquium. The interdisciplinary discussion has shown us that we must question everything and be prepared to evaluate traditional truths in a new light. This fact is certainly true in regard to physics and its recent discoveries: while physicists themselves are painfully aware of the limits of their knowledge, the general public still tends to see physical discoveries in absolute terms. For philosophers our colloquium may act as a reminder that, contrary to the claims of positivists, logical empiricists, and many other of today's philosophical schools, metaphysics is not an out-dated field of philosophical inquiry. On the contrary, it may offer valuable insights for the philosophy of contemporary physics. As for theologians, it is time for them to believe that they can contribute to the cosmological discussion not only by offering a historical perspective to the shared past but also by bringing in the intuitive theological notion that the cosmos is not a product of random occurrence but that all beings have a reason for their existence. One may conclude that an interdisciplinary dialogue among theologians, philosophers, and physicists concerning the future and the past of cosmological inquiry can teach each of these disciplines not only about each other, but also about themselves and their specific subjects

Acknowledgements

There are many to whom I am very grateful for their cooperation and help in arranging the colloquium. Antje Jackelén, the director of the CTNS Science & Religion Course Program in Europe suggested in October 1999 that we organize a colloquium on historical and contemporary cosmology. Tarja Kallio-Tamminen and Atso Eerikäinen, members of my research project, have assisted in many ways in the planning and arranging of the colloquium. Suvielise Nurmi, the secretary of the colloquium, has taken care of numerous practical arrangements. Maiju Lehmijoki-Gardner and Christopher Gardner translated and corrected papers and other material related to the colloquium. Lassi Larjo has assisted in the publication of this volume as a redactor and David Jenkins has helped to polish the language of several articles. Finally, I would like to thank Kari Enqvist who was very generous with his time and in sharing his expertise in physics with the members of the project.

2. Causality and Cosmology

The Arabic Debate

Taneli Kukkonen (Helsinki)

1. Introduction

This paper intends to introduce one of the main themes of this collection: namely, the question of causality in its cosmological setting. More specifically, I wish to examine the interplay between notions of divine and natural causality. A commonplace of Western theology has been to stress the transcendence of God. Whatever we take nature and natural causality to be, this does not characterise what God is like and how God acts. Rather, God is the 'cause of causes', the ultimate grounds for the world's existence, who allows for the efficacy of secondary causes. This was the view of, e.g., Thomas Aquinas; and I assume it is the most influential one.[1]

The contention is, of course, nothing new. Neither is it in any way peculiar to Christianity. Lloyd Gerson has persuasively argued that the major part of Greek philosophy is best understood as a continued discourse on a special kind of natural theology. According to Gerson, when the Greek philosophers attempted to formulate how the ultimate *arkhê* of things differs from regular causes encountered in the course of nature, what they really tried to do was to argue why the First Principle or *arkhê* could be said to be categorically different from regular causes, grounding all, yet itself remaining uncaused.[2]

The fortunes of the Greeks within the Christian world have been well charted. Here I propose to throw light on a different line of influence. I would like to examine the evolution of philosophical notions of causality in the Islamic context during the 9th through the 12th centuries. The debate between the Muslim philosopher

---

1 In the Latin tradition the division between natural and supernatural causes can be traced back to Calcidius, the 4th century commentator on Plato: see J.H. Waszink (ed.), *Timaeus a Calcidio translatus commentarioque instructus,* cap. 23 (London ²1975), 73–74. For Muslim affirmations that God is the 'cause of causes' see Abû al-Ma'âlî al-Juwaynî, *Kitâb al-irshâd* ('Book of Guidance') (Cairo 1950), 83; and Abû Hâmid al-Ghazâlî, *Al-maqsad al-asnâ fî sharh asmâ' Allâh al-husnâ* ('Exposition on the Beautiful Names of God') (Beirut 1986), 98 (English translation in: *The Ninety-Nine Beautiful Names of God,* tr. D.B. Burrell & N. Zaher (Cambridge 1992), pagination to the Arabic indicated).
2 L.P. Gerson, *God and Greek Philosophy* (London 1990).

19

Averroes (1126–1198 C.E.) and the renowned theologian al-Ghazâlî (1058–1111) in the latter's 'Incoherence of the Philosophers' (*Tahâfut al-falâsifa*, 1095) and the former's response, the 'Incoherence of the Incoherence' (*Tahâfut al-tahâfut*, 1180) can help to spotlight the salient points in the Arabic discussion.[3] The main problem is whether an eternal world can meaningfully be said to be the subject of God's actions, or whether a contingent and limited creation better corresponds to the notion of an all-powerful, all-knowing and all-benevolent Creator. Averroes' and al-Ghazâlî's exchange touches on each of the themes outlined in the heading of this colloquium: taking these in order, infinity, causality, and determinism. I would therefore like to take this opportunity to introduce some of the problems historically associated with these issues.

A brief synopsis of what I am going to do is in order. After this introduction (1) I shall first trace the historical roots of the connection traditionally made between causation and creation; (2) that is, the idea that the paradigmatic case of one thing causing another is when something is brought into existence. (3) This account stands in stark contrast to the Islamic philosophers' understanding of the Aristotelian universe as eternal. (4) The philosophers and theologians interpret in very different ways the common principle that an infinite regress of causes is impossible. (5) A look at the famed occasionalism debate in the *Tahâfut* goes to show just how far apart the theological and philosophical conceptions of the universe are in the Islamic context. (6) I will argue that the difference between them goes back to varying intepretations of the familiar 'Principle of Sufficient Reason'. (7) This brings us to the issue of determinism. Time, tense, and truth-values converge in these discussions.

2. *Causes and Origins*

Throughout the Western tradition, divine causality has been linked to the mystery surrounding the universe's origins. It is not altogether clear why this should form the main focus of attention. Mystics and theologians of every persuasion have time and again reminded us that we can – maybe even should – look for signs of God's hand in the microscopic details of existence as much as in its grand outlines.[4] Still, there is a persistent interest in ultimate origins which is not easily shaken,

---

3   The edition used for the *Tahâfut al-tahâfut* (hereafter, TT) is by M. Bouyges (Beirut 1930). For al-Ghazâlî's *Tahâfut al-falâsifa* (hereafter, TF) I use the splendid new *The Incoherence of the Philosophers. A parallel English-Arabic text*, ed. and trans. M.E. Marmura (Provo, UT 1997). All translations from the *Tahâfut al-falâsifa* are Marmura's; those from the *Tahâfut al-tahâfut* are my own.
4   Al-Ghazâlî, who in his later years was equal parts theologian and mystic, routinely conjoins the micro- and macroscopic: see, e.g., *Al-maqsad al-asnâ*, 79–82 and 105–8 Shehadi.

as shown already by the amount of popular contemporary literature devoted to Big Bang Cosmology.

Our abiding fascination with absolute beginnings is effortlessly traceable to the two mainsprings of Western thought: the shared Judaic heritage of the *Genesis* account and the stratum of Platonic ideas at the core of our intellectual tradition. The former is familiar enough: suffice it to say that the *Genesis* creation stories (of which there are two) admit in themselves of diverse interpretations, but the prevailing opinion among the children of Abraham came to be that revelation discloses a divine creation out of nothing in some limited time in the past.[5] But the second heritage, the Platonic one, calls for immediate comment.

A popular way of setting up a contrast between the Hebrew and Greek elements in our intellectual heritage has been to suggest that Semitic thought is in some way voluntarist, historically oriented, and personalistic, whereas Greek thought fixes its sights on the rational, the abiding, and the impersonal. Whatever the other merits of this generalisation, it is worth noting that on the question of creation and causation the Platonic tradition, at least, occupies a rather more ambivalent position than this caricature would suggest. True enough, Platonic thought orients itself towards an immutable formal level of being, and it contrasts this level with the sensory and transitive realm of becoming. But precisely because it does so, the corporeal and sensible world can come to be viewed as being generated. The Arabs would learn of Plato's 'likely tale' of cosmology, the *Timaeus,* from Galen (d. *c.* 199 C.E.) in a paraphrase now only extant in the Arabic translation. As Galen recalls the first and most fundamental query in that work (27d),

> 'What is that which always is, having no birth; and what is that which is always being born but which at no time exists?' The answer to this question is evident to one who is versed in the various works of Plato: for he distinguishes between intellectually apprehensible substances, which are not in bodies, and those that are sensible. And the latter Plato customarily calls 'becoming', rather than substances...[6]

Galen's exposition goes far beyond the needs of mere *Timaeus* exegesis. For Galen, it is evident that only immaterial Ideas are true substances; by contrast, sensible

---

5   For the Christian context see G. May, *Schöpfung aus dem Nichts: Die Entstehung der Lehre von der Creatio ex nihilo* (Berlin 1978). See also H.A. Wolfson, 'The Meaning of ex nihilo in the Church Fathers, Arabic and Hebrew Philosophy, and St. Thomas,' in: *Studies in the History of Philosophy and Religion* (Cambridge, MA 1973), vol. 1, 207–21; and 'The Twice-Revealed Averroes,' in: Wolfson, *Studies,* 371–401, 373–78.

6   P. Kraus & R. Walzer (eds.), *Galeni Compendium Timaei Platonis* (London 1951), Arabic text, 4.2–6; on the being-becoming distinction see the papers by R.G. Turnbull, G. Fine, M. Frede and A. Code in: J. Annas (ed.), *Oxford Studies in Ancient Philosophy Supplementary Volume* (Oxford 1988), 1–60.

and corporeal bodies are only generated, always a-borning, never existent. Galen goes on to remark that the above classification is intended as exclusive and exhaustive: all things either always are and are never born or else never are but are only generated. Another crucial observation follows. Plato (Timaeus) says that

> 'Everything which is generated must necessarily do so from a cause': and he does so without giving a demonstration for it. This is because this is one of those things that are evident to reason. For if there is anything whatsoever which is continuedly in one and the same state, neither generating nor corrupting, then that thing has no generative cause [^{c}illa mukawwina]. Every generated thing, on the other hand, has an efficient cause [^{c}illa fā^{c}ila]; and everything that is generated has an efficient cause present at the time [of its generation].[7]

So generation entails efficient causation and an efficient cause, simultaneous with the individual birth as well as preceding it: this fact is viewed as self-evident, not requiring proof. Add to this one of two premises, either

(a) an infinite regress of causes is impossible, or
(b) whatever is connected with or consists of generated things is itself generated,

or indeed both, and we arrive at the conclusion that the universe must have had a temporal beginning.[8] Furthermore, if nothing comes from nothing, then we also need an utterly transcendent First Cause, to set everything in existence and in motion. This constitutes the cosmological proof from creation for the existence of God. It is, in fact, what William Lane Craig in our time would call the *kalām* cosmological argument. I believe Craig is right to emphasise the Muslim contribution in its development.[9]

Going back to our *Timaeus* paraphrase, we find Galen contending that the narrator had his sights on such a proof all along: 'as concerns the world's birth, whether it never began or whether it had a beginning, he decides on the matter

---

7   *Galeni Compendium Timaei Platonis*, 4.14–19.
8   For (a) see Averroes' *Questions in Physics*, q. 4, ed. and trans. H.T. Goldstein (Dordrecht 1991), 8–9 (similarly q. 7; 19–20 Goldstein and TT, 171–72); for (b), *Kashf ^{c}an manāhij al-adilla fī ^{c}aqā'id al-milla* ('Exposition of the Methods to Prove the Truths of Religion'), ed. M.^{c}Â. Jâbirî (Beirut 1998), 104. In the *Tahâfut al-tahâfut*, 220–21, Averroes mentions both (a) and (b). For the history of the arguments see H.A. Davidson, *Proofs for Eternity, Creation, and the Existence of God in Medieval Islamic and Jewish Philosophy* (Oxford 1987), 117–46.
9   On the irreducibility of ex nihilo nihil fit see C.D. Broad's comments in: W.L. Craig & Q. Smith, *Theism, Atheism, and Big Bang Cosmology* (Oxford 1993), 57–63; cf. however Smith's comments at 178–91.

in what follows, and tells us that its birth had a beginning.' About this birth and about its Cause it is difficult to say anything more, since it belongs to another category of causes altogether from what we observe in nature. Certainly a pious sentiment: no wonder that the Muslims would hail Timaeus as one of their own, a *mutakallim* or 'theologian'.[10] A Philo or a Justin might claim that Plato learned, or else borrowed and stole from Moses: but for the Muslim theologians such a question would be irrelevant. That the contingent world neaeds a Creator was considered to be self-evident to any rational observer. The generally held view was that this much 'natural theology' came naturally to anybody who contemplated the existence of the world. The argument from creation was thus a staple of Islamic rational theology: it was considered to be the first step of religious guidance.[11]

3. *Eternity: No Room for Creation*

The Arabs, then, right or wrong, by and large knew Plato as a creationist of sorts.[12] Not so with Aristotle. Aristotle had unmistakably insisted on the point that neither motion, nor time, nor worldly process has beginning or end; and the Arabs were well acquainted with the Stagirite's natural philosophy, as his was the main form of worldly wisdom (*hikma*) they inherited from the late ancient cultural scene. The overarching superstructures of the Aristotelian universe (the celestial realm, prime matter, the unmoved movers) were clearly meant to be permanent and indestructible. Even discounting these, Aristotle in several places indicated that sublunary organisms, too are quite capable of endlessly reproducing themselves.[13]

---

10 See *Compendium Timaei Platonis*, 4.20–5.3 Kraus & Walzer; cf. L.E. Goodman, *Avicenna* (London 1992), 50ff.
11 See al-Juwaynî, *Irshâd*, 3 Mûsâ & ʿAbd al-Hamîd; Averroes, *Kashf*, 102–4 Jâbirî; and al-Ghazâlî, *Al-maqsad al-asnâ*, 158.2–4 Shehadi, as well as *Al-iqtisâd fî al-iʿtiqâd* ('The Golden Mean in Belief'), ed. M. Cubukcu & K. Atay (Ankara 1962), 32–33 and ff. For a parallel in Ghazâlî's famous 'Jerusalem Letter' see A.L. Tibâwî, 'Al-Ghazâlî's Sojourn in Damascus and Jerusalem,' *Islamic Quarterly* 9 (1965), 65–122, at 80.24–29. For a study, cf. L.E. Goodman, 'Ghazâlî's Argument from Creation,' *International Journal of Middle East Studies* 2 (1971), 67–85 and 168–88.
12 Controversy about Plato's cosmogony has raged for 23 centuries. For the ancient discussions see the magisterial M. Baltes, *Die Weltentstehung des platonischen Timaios nach den Antiken interpreten*, 2 vols. (Leiden 1976–8). In favour of a non-literal interpretation see M. Baltes, 'Legomen (Platon, Tim. 28B7): Ist die Welt real entstanden oder nicht?', in: K.A. Algra et al. (eds.), *Polyhistor* (Leiden 1996), 79–96. The Muslims certainly knew of the controversy: see al-Ghazâlî, TF, 1; Marmura (ed.), 12.6–13. They also knew of the doctrine of 'disordered motions' preceding the orderly cosmos: cf. Averroes, *Tafsîr mâ baʿd al-tabîʿa* ('The Commentary on the Metaphysics'), ed. M. Bouyges (Beirut 1938–52), bk. 12, comm. 30 (3:1570–73).
13 On the world's incorruptibility see the author's 'Al-Ghazâlî's and Averroes on the End of All Things'. Forthcoming in *Medieval Philosophy and Theology*. For the species see J.G. Lennox,

This last view in particular was at times given an unabashedly materialist interpretation in later Greek philosophy, notably by the Stoics. Perhaps this is what led al-Ghazâlî to describe materialism as the belief that 'everlastingly animals have come from seed and seed from animals; thus it was and thus it ever will be.'[14] 'Materialism' in Arabic translates as 'eternalism' (*al-dahrîya*): the two expressions are equivalent, and had been so for a long time. So al-Ghazâlî's indictment was nothing new. What al-Ghazâlî did more clearly than anyone before him was to lay out the strictures of the Aristotelian eternalist scheme and point to how these inevitably lead towards irreligion.

The challenge is nowhere so clearly formulated as in al-Ghazâlî's opening statement in the Tenth Discussion of *The Incoherence of the philosophers,* which attempts to prove 'their inability to show that the world has a maker and a cause'. The contrast with the theologians is marked. When the latter 'maintain that every body is temporally originated because it is never devoid of temporal events' and from this infer that the world 'needs a maker and a cause, their doctrine is intelligible'.

But as for you [philosophers], what is there to prevent you from [upholding] the doctrine of the materialists – namely, that the world is eternal, that no body in the world is originated and no body annihilated, but [that] what occurs temporally is forms and accidents? For [according to this doctrine] the bodies consist of the heavens, which are eternal, [and] the four elements constituting the stuff of the sublunar sphere. The bodies and materials [of the latter] are [likewise] eternal. It is only that the forms, through mixtures and transformations, undergo successive changes over [these bodies]; the human and vegetative souls come into temporal existence. The causes of [all] these [temporal] events terminate in the circular motion, the circular motion being eternal, its source an eternal soul of the heavens. Hence, [according to the materialists] there is no cause for the world and no maker of its bodies, but it continues eternally to be in the manner that it is, without a cause (I mean [without a cause of its] bodies). What, then, is the sense of their saying that these bodies come into being through a cause, when [such bodies] are eternal?[15]

'Are Aristotelian Species Eternal?', in: A. Gotthelf (ed.), *Aristotle on Nature and Living Things* (Bristol 1985), 67–94.

14 Al-Ghazâlî, *Al-munqidh min al-dalâl* ('Deliverer from Error'), ed. J. Salîba & K. ᶜAyyâd (Beirut), 96; translation by W.M. Watt in: *The Faith and Practice of al-Ghazâli* (Oxford 1993), 30. Al-Ghazâlî sets up a threefold division between 'materialists' (al-dahrîyûna), 'natural philosophers' (al-tabîᶜîyûna), and 'metaphysicians' or 'theists' (al-ilâhîyûna): he concludes that even the latter (who included Socrates, Plato, and Aristotle) could not rid philosophy of its irreligious tendencies. See *Munqidh,* 96–99 Salîba & ᶜAyyâd.

15 TF, 125.

Al-Ghazâlî takes care to paint a picture of the philosophers' universe as a closed system that is 'designed to feed itself', as Plato had described the process.[16] The world is a *perpetuum mobile,* forever rolling onward, each part seamlessly interlocking with all others. This leaves us with two questions.

One: On this kind of view, *could* there be a temporal beginning to the universe? The Aristotelian philosopher would have to answer in the negative. What Aristotle had proven was precisely that there *cannot* be an absolutely first event in the temporal sense. Some prior motion precedes every conceivable motion, as is stated in the beginning of *Physics* 8. But this means that the theologians' understanding of the cosmological argument is invalid.[17] This realisation brings us to our second question: Can an eternal world have a First Cause at all? This in turn the theologians would deny, even if the philosophers might wish to argue otherwise. If the dichotomy posed by Timaeus holds true – if unbegotten things have no generating cause – then this assertion seems almost a foregone conclusion. As Averroes would observe, 'It was difficult for the Muslims to call both God and world eternal, since they had no other understanding of the eternal than that which has no cause'.[18] Al-Ghazâlî says as much in the *Incoherence of the philosophers* when he states that

> The meaning of 'act' is the bringing forth of things from nonexistence to existence by creating it. But this is inconceivable of the pre-eternal, since what [already] exists cannot be brought into existence. Hence, the condition of the act [to be something enacted] is for it to be temporally created. But the world, according to [the philosophers], is pre-eternal. How could it then be the act of God?[19]

One must note that this is a very specific take on the nature of causality. For the idea that causation equals origination is not Aristotelian. Aristotle recognises four different kinds of causes: material, efficient, formal, and final. None of these exactly corresponds to creation. If the theologians would note this, so would the philosophers. In fact, the Neoplatonist Avicenna (980–1037) exploited this very gap in the Aristotelian scheme in developing his own peculiar metaphysics. According to Avicenna, if God is the cause of the world, He operates as an efficient cause or 'agent' [$fâ^cil$]: but first and foremost in the special sense of bestowing

---

16 Plato, *Timaeus,* 33d.
17 As Averroes states with some acerbity in the *Commentary on Aristotle's Physics,* 8, comm. 47; preserved in the Renaissance Latin *Aristotelis opera cum Averrois commentariis* (Repr. Frankfurt am Main 1962), 4:f.388E–L. (This series will hereafter be abbreviated 'AOAC'.)
18 TT, 124.11–12.
19 TF, 61.4–7.

existence. In physics, efficient causality 'produces motion in something other than itself', as Aristotle had indicated. But in the more fundamental, metaphysical form of agency, something gives *existence* to something other than itself. And this is the kind of action which we should most properly predicate of God.[20] God is the eternal source of the world's existence. It is Avicenna that al-Ghazâlî is following here: both follow what are at heart Platonic lines of investigation.[21] Al-Ghazâlî only lays emphasis on the temporal aspect of divine action and thereby upsets Avicenna's scheme of an eternal emanation.

To Averroes' mind, Avicenna's vulnerability to al-Ghazâlî's criticisms serves to prove the deficiency of the entire line of reasoning employed by Avicenna and the emanationists. Averroes is of the opinion that Aristotle had sufficiently demonstrated the place and mode of causality that God holds over the world. Aristotle postulated at the top of his cosmology a series of immaterial movers and proceeded to call the first (or last) of these 'God'.[22] The First Cause is the First Mover, the mover of the outermost sphere and the sustainer of the world's diverse motions. Since the world 'consists of motion', keeping the world in motion is equivalent to keeping it in existence. In addition to this, God is the world's final cause: He is that toward which all being strives and through which everything gains its perfection. By Averroes' count, this accounts for three of Aristotle's classification of the four causes: the formal, the final, and the efficient. The fact that God is not the material cause of the world's existence hardly entails any imperfection.[23]

So we have two fundamentally different conceptions of how God is the First Cause of the world. For the Ash^{C}arite theologians, God stands at the world's temporal beginning; for the Aristotelians, God is the eternally active *primus motor,* moving the outermost heavenly sphere and thereby directing the rest of the universe.

---

20 Avicenna, *Shifâ', Al-samâ^C al-tabî^{C}î'* ('The Healing: Physics'), 1.10; ed. S. Zayed, revised by I. Madkûr (Cairo 1983), 48.13 and 49.11–12; followed by *Shifâ, al-Ilahîyât* ('The Healing: Metaphysics'), 6.1; ed. M.Y. Mûsâ et al., revised by I. Madkûr, 2 vols (Cairo 1960), 2:257. (Hereafter abbr. 'Ilâhîyât'.) Cf., further M.E. Marmura, 'The Metaphysics of Efficient Causality in Avicenna (Ibn Sina),' in: M.E. Marmura (ed.), *Islamic Theology and Philosophy* (Albany, NY 1984), 172–87.
21 Avicenna calls eternal emanation 'true origination' (ibdâ^C): see, Ilahîyât, 6.2; 2:266–67 Madhkûr. Cf. F. Rahman, 'Ibn Sina's Theory of the God-World Relationship,' in: B. McGinn & D.B. Burrell (eds.), *God and Creation* (Notre Dame, IN 1990), 38–51. This interpretation of divine agency goes back at least to Proclus (d. 485 C.E.): see his *In Platonis Timaeum commentaria,* ed. E. Diehl, 3 vols. (Leipzig 1903–6), 1:261–62.
22 On the celestial hierarchies see Averroes, TT, 191–94, 214–18, and 231–34. See further Tafsîr, 12, comm. 44; 3:1646–53 Bouyges (in the Latin, AOAC 8:f.327D–328G).
23 See, e.g., TT, 162, 163–65, 172. For the equation of motion and existence, cf. in addition *De substantia orbis,* 4; *Averroes' De Substantia Orbis* ed. A. Hyman (Cambridge, MA 1986), Hebrew text, ll. 30–38, English translation 117 (the Latin in AOAC 9:f.10l). For commentary, B.S. Kogan, *Averroes and the Metaphysics of Causation* (Albany, NY 1985), 203–21.

We might take these conceptions to represent the *creatio* and *conservatio* traditions respectively; and there is much to recommend this interpretation. However, matters are not quite that simple. Avicenna and Averroes both maintain that their own model credits God with true creation, as the other does not.[24] Also, there is a great deal of conservation going on in the theologians' model, as we shall see.

4. *Finitude and Infinity*

Let us return to the first of the above two questions. Is a first moment and a first movement possible? The Aristotelian philosophers denied this on the basis of diverse conceptual considerations. To take one example, 'the moment' itself, by Aristotle's definition, represents the borderline between the future and the past. Thus, whatever moment we take, it necessarily has a time preceding and a time following. And since time is but the 'measure of motion with regard to prior and posterior' (thus again Aristotle), motion, too, is eternal.[25] Other similar proofs of the philosophers argue for the pre-existence of motion, of potency, and of matter as the carrier of that potency.

Against all these alleged proofs al-Ghazâlî presents a single comprehensive counterargument. He asks: is a temporal creation really logically contradictory? If it is, then the philosophers must produce a strict demonstration to prove it. But al-Ghazâlî feels confident the philosophers cannot accomplish this.[26] If he is correct, then it seems the philosophers' so-called 'proofs' only impose unwarranted restrictions on the free agency of God. God operates in a way different from natural causes. As the Christian Monophysite John Philoponus (490–574 C.E.) had put the case, if God 'produces in a way similar to nature, then he does not differ from nature': and if he does not differ from nature, how then is He superior to it? Rather, since God's power over nature is absolute, He has generated motion, time, matter, form and potency all simultaneously from nothing, before which there was absolutely nothing.[27] Natural laws and definitions may have a place

---

24 Averroes came to believe that Avicenna's scheme leaves the door open for atheism and materialism: TT, 419–23. See also C. Steel & G. Goldentops, 'An Unknown Treatise of Averroes against the Avicennians on the First Cause. Edition and Translation,' *Recherches de Théologie et Philosophie médiévales* 54 (1997), 86–135 and Davidson, *Proofs*, 312–18. For Avicenna's polemics against the Peripatetic model, see the preserved comments on *Metaphysics Lambda* published in ᶜA. Badawî (ed.), *Aristû ᶜinda al-ᶜarab* (Cairo 1947), 23–26.
25 See e.g. TT, 76.10–15 together with 65.4–6 (see *Phys.* 4.13, 222a10–12 and b1–2). For the converse proof – from motion to time – see *Phys.* 8.1, 251a7–b28 and TT, 84–86.
26 TF, 15.1–5 and 17.6–15.
27 Philoponus, apud Simplicium, *In Aristotelis Physicorum libros quattuor posteriores commentaria* ed. H. Diels (Berlin 1895), 1150.21–25, 1141.11–30, and 1142.1–28. All passages are available in English in *Philoponus Against Aristotle, on the Eternity of the World*, trans. C. Wildberg

inside of nature; but they certainly do not impose restrictions on the transcendent activity of God. We recognise this as a stock proof for the possibility of miracles, divine intervention, and *creatio ex nihilo* even to this day.

Al-Ghazâlî goes on to make a point about the Aristotelians' supposed proof for the eternity of time. The mind is unable to imagine a limit to time without imagining a time that would precede it, this is true. But the mind is likewise incapable of picturing the world's outermost edges in *space*, beyond which there is nothing. Nevertheless, the latter (the spatial finitude of the world) is standard philosophical doctrine. What is there to distinguish between the two cases? In truth, both the world's spatial and its temporal co-ordinates have limits, even if our imagination feels compelled to reach beyond them.[28]

So much for the possibility of an absolute beginning. Al-Ghazâlî, again following Philoponus, believes he can go further and demonstrate that a first moment and a first motion is actually *necessary* – on the philosophers' own assumptions, no less. The argument is based on Aristotle's denial of an actual and traversed infinite. In the *Incoherence of the philosophers*, several impossibilities resulting from an infinite past are pointed out. If time has no beginning, then the past revolutions of the sun are infinite; but so are the revolutions of Saturn, and yet the one should be as much as 30 times the other in quantity. Also, past revolutions would be either even or odd; but then with the passage of one day the infinite would have been increased, as even would become odd and odd, even. As Averroes notes, all of these are appeals to the principle that one infinite cannot be greater than another. This claim in turn rests on Aristotle's statement that 'the same thing cannot be many infinites'.[29]

All of these problems are corollaries to the more fundamental argument Averroes sees lurking behind al-Ghazâlî's statements. The real creationist challenge, the one Averroes regards as the most serious difficulty the theologians had ever raised against the philosophers, runs as follows: 'If the movements which have occurred in past time are infinite, then at the present moment no movement can exist or occur, unless an infinite [amount of] movements has been completed before it.' But as Philoponus had pointed out centuries earlier, this should be impossible: for

---

(London 1987). Cp. al-Ghazâlî, TF, 20.19–20, 31.12–19, and 42. For other medieval instances of this kind of approach see Davidson, *Proofs*, 30–33.

28 TF, 31.12–19 and 32.13–33.17. This point, too, derives from Philoponus. See the comment preserved under the name 'Yahyâ', i.e. John, in the Arabic version of Aristotle's *Physics: Aristûtâlis, Al-tabîʿa*, ed. ʿA. Badawî, 2 vols. (Cairo 1964–5), 816.14–17.

29 TF, 18.9–19.2 (cf. Philoponus, apud Simplicium, *In Aristotelis Physicorum libros*, 1179.15–24 Diels) and TT, 19.10–12. For the Aristotelian background, *Phys.* 3.5, 204a25 and *Met.* 11.10, 1066b15: see the comments in Philoponus, *In Aristotelis Physicorum libros tres priores commentaria*, ed. H. Vitelli (Berlin 1887), 468.1–2 and Averroes, *Commentary on Aristotle's Physics*, 3, comm. 37; AOAC, 4:f.102C.

by Aristotle's own testimony the infinite is by its nature untraversable.[30] A closely related argument which we have already encountered argues for the impossibility of an infinite regress of causes. Since in an infinite series of causes the First (=the last) is unattainable, the whole series remains ungrounded: it hangs suspended on nothing, so to say, and so cannot actually have come to be.[31]

Al-Ghazâlî develops a shrewd variation on this last proof. Let us for the sake of argument suppose that an infinite regress of causes *is* possible. There is then nothing to prevent us from assuming that the world is ungenerated and imperishable, like the philosophers say. But we must then also admit that the world as a whole is uncaused, the same way it is, as a whole, atemporal and, again as a whole, necessary. Why? Because

> ...what is true of those [beings] that have beginnings is applicable to the individual units but not true of the totality. Similarly, it can be said about each individual unit that it has a cause, but it is not said that the totality has a cause. Not everything that is true of the individual units is true of the totality... Every place on earth that we specify is lit by the sun during the day [and] becomes dark at night, and each [of these events] comes into existence after not being – that is, it has a beginning. But the totality for them [, the philosophers,] is that which has no beginning. Hence, it has become evident that whoever allows the possibility of events that have no beginning... is unable to deny causes that are infinite. From this it comes about that they have no way of reaching [the point] of affirming the First Principle...[32]

So the philosophers' theory is in actual fact tantamount to atheism, even if they might want to claim otherwise. Al-Ghazâlî inventively pairs off cases in which the philosophers would say that what is true of the part is not true of the whole. He additionally points out that in the philosophers' universe, there are several infinite sets, such as, e.g., past heavenly rotations and Avicenna's immortal souls. Why then should there not be also an infinite regress of causes, a circumstance which, however, obviates the need for a First Cause? The philosophers' discrimination here is purely arbitrary.[33]

---

30 TT, 20.1–4; cf. Philoponus, *De aeternitate mundi contra Proclum*, 1, ed. H. Rabe (Leipzig 1899), 9.14–10.21, drawing on Aristotle, *Phys.* 3.4, 204a2–4.
31 Cf. Simplicius, *In Aristotelis Physicorum libros*, 1178.20–35 Diels. The argument depends on Aristotle, *Met.* 2.2, 994a1–20, where the infinite transformation of the elements into other elements and an infinity of desired things is blocked out.
32 TF, 83.4–12.
33 TF, 80.19 and 83.12. On the infinite number of souls see TF, 19 and 81–84; cf., M.E. Marmura, 'Avicenna and the Problem of the Infinite Number of Souls,' *Mediaeval Studies* 22 (1960), 232–39.

It is here that Averroes sees the chance to intervene. It is not true that there is no difference between temporally precedent and temporally simultaneous causes: on the contrary, their difference is all-important. Consequent causes are *accidental;* simultaneous chains of movers are *essential.* The latter are the true explanatory causes and thus must be finite. The latter may well be without beginning or end. In fact this is required, since it is precisely that all of the components are the same and present throughout the whole of history that makes the essential moving chain omnitemporally invariant, hence universally true.[34] In Averroes' estimation, the division between accidental and essential causes is the real reason Aristotle, in the course of his proof for God, reminds us that man and the sun together beget man. The ancestor is only instrumental in the production of a new existent; the pieces of the continuous chain of movers are the real agents conveying God's providential care throughout all time.[35]

This last feature merits an explanation. In the medieval Arabic understanding of Aristotle, the movements of the sublunary particulars are governed by the celestial spheres. The motions of the elements result from the attraction and repulsion they assert, and the generation of life comes from the emission of celestial heat.[36] The celestial movements, so Averroes contends, in turn gain their order and cohesion from the first motion (that of the outermost sphere) and the First Mover. This is why Averroes feels entitled to say that God, even in moving infinitely and, so to say, indiscriminately, extends His providence to all creatures.[37] If not for the unifying function of God's Wisdom and the motion which He initiates (both of which all worldly intellections and motions strive to imitate) all the world would fall into ruin and disrepair. As it stands, the world traces and infinitely retraces the finite cycle in which the chain of movers reproduces itself.[38]

---

34 TF, 20–23. See also Questions, q. 4; Goldstein ed., 8–9 and Long Commentary on the Physics, 8, comm. 15; AOAC 4:f. 349rbEff. On eternity and necessity in Averroes' thought see the author's 'Possible Worlds in the Tahâfut al-tahâfut. Averroes on Plenitude and Possibility,' *Journal of the History of Philosophy* 38/3 (2000), 329–347.
35 TT, 268–69. On 'the sun and the man' see Aristotle, *Met.* 12.5, 1071a14–16 (cp. *Phys.* 2.2, 194b14); the passage adds 'the ecliptic', for whose significance see the next note below. Averroes claims that the distinction solves a long-standing problem regarding the interpretation of *Physics* 8: see his Commentary on the Physics, 8, comm. 1; AOAC, 4:f.338F and ff.
36 The theory was developed in consecutive stages from scattered hints in Aristotle (cf. *De gen. et corr.* 2.10, 336a15–337a33; *Met.* 12.6, 1072a9–18; *Meteor.* 1.9; GA 4.10) by Alexander of Aphrodisias, by al-Kindî's circle, and by al-Fârâbî. For these developments see S. Fazzo & H. Wiesner, 'Alexander of Aphrodisias in the Kindî-Circle and in al-Kindî's Cosmology,' *Arabic Sciences and Philosophy* 3 (1993), 119–53 and Th.-A. Druart, 'Al-Fârâbî's Causation of the Heavenly Bodies,' in: P. Morewedge (ed.), *Islamic Philosophy and Mysticism* (Delmar, NY 1981), 35–45.
37 TT, 185–94, 216–20, 229–34, 261.
38 See e.g. TT, 163–65, 168, and 229–30; cf. further the author's forthcoming 'Al-Ghazâlî and Averroes on the End of All Things', sec. 3 and *Talkhîs fî al-kawn wa al-fasâd* ('Paraphrase of [Aristotle's]

We may make brief note of the fact that as far as cosmological infinity is concerned, what Averroes considers crucial is the finitude related to the world's *accountability*. For the world to be rationally understandable, it must be fully comprehensible in a finite number of steps. This is why the world can be infinitely divisible but never infinitely divided and why it can extend *ad infinitum* temporally but never hold an actually infinite intelligible content. This perceived relation between finite knowability and 'ultimate grounds' has survived to this day in the dream that a comprehensive 'Theory of Everything' would contain in compressed form the entire universal order and disclose its inherent necessity.[39]

5. *Occasionalism and Natural Laws*

Piecing together what we have learned thus far, both the Islamic philosophers and the theologians believed that God is demonstrably the First Cause of the world. In arguing for their view, both parties took recourse in two shared axiomatic principles: that every event has a cause and that an infinite regress of causes is impossible.[40] We have examined the varying ways in which the philosophers and theologians interpreted the second of these notions. Let us now move on to the first, the principle of causality itself.

The philosophical worldview paints a picture of a complex assembly of entities both great and small working in tandem. Each moves some things and is in turn moved by others, each works towards actualising its attendant potencies in imitation of the eternal actuality of God. This is not how the Islamic theologians' world was arranged. According to the metaphysical doctrine which had become prevalent among the theologians by the 10th century, the only real operative causal agent is God. The world consists of bare, featureless atoms, the minimal building

---

De generatione et corruptione'), ed. J. al-Dîn al-ᶜAlawî (Beirut 1995), 119ff., as well as *Risâla al-kawn wa al-fasâd* ('The Epitome of [Aristotle's] De generatione et corruptione'), ed. R. ᶜAjam, in Rasâ'il Ibn Rushd al-falsafîya, 6 vols (Beirut 1994), 3:113ff. Both passages are available in English translation in *Averroes on Aristotle's De Generation et Corruptione: Middle Commentary and Epitome,* tr. S. Kurland (Cambridge, MA 1958), 97ff. and 128ff.

It would seem that Averroes successfully deflects the charge of an infinite causal regress. (The question might not even bother a modern philosopher: see R. Sorabji, *Time, Creation, and the Continuum* (London 1983), 230–31.) However, at least two problems remain. The charge of a traversed infinite still stands (*vide* al-Ghazâlî's points at TF, 19.10–13); also, there are inconsistencies in Averroes' account of rectilinear vs. cyclical causes (Davidson, *Proofs,* 132–33, n. 128).

39 I have elsewhere commented on the conceptual presuppositions lying at the heart of this conception: see 'Plenitude, Possibility, and the Limits of Reason: A Medieval Arabic Debate on the Metaphysics of Nature' *Journal of the History of Ideas* 61/4 (2000), 539–560.

40 Aquinas, among others, notes that these two principles lie at the heart of the Aristotelian proof from motion: see *Summa contra Gentiles,* I, 13.

blocks of physical reality which the theologians proceeded to call 'substances' (*jawâhir*). To these substances God at each moment of time attaches the qualities (or 'accidents', ^{c}*arâd*) He decides upon. God's power over the atoms is free and absolute. Thus God can make any saint into a sinner and extend His grace to the unrepentant killer. From the physical aspect, it is clear that by the tenets of this view miracles and prophetic signs are perfectly possible. The secondary causality assigned to natural agents, meanwhile, is merely an illusion, a figment of our imagination. In truth, the world is unable even to sustain its own existence: at the 'end' of each discrete moment of time all features are lost to the atoms, and only God's grace reinstates them in the manner we see. This metaphysical scheme is known as occasionalism, since in it the world is fashioned out of subsequent 'occasions'. Another term used has been 'Islamic atomism'. This emphasises the discontinuous manner in which the Ash^{c}arite universe is put together, both spatially and temporally.[41]

The extent of al-Ghazâlî's adherence to occasionalism is at the present time a matter of some controversy.[42] I will not go into the details of the debate, since it does not serve our purpose. This much is undisputed. In the causality debate on the pages of the *Incoherence of the philosophers* al-Ghazâlî presents for the reader's consideration two rival theories. One scheme is purely occasionalist; the other recommends what we might call cautious naturalism.[43] And al-Ghazâlî attempts to show that whichever theory one prefers, one must allow for the possibility of miracles and divine intervention.

As regards the occasionalist theory which denies secondary causality altogether, al-Ghazâlî's line of argumentation is as simple as it is effective. Once again al-Ghazâlî draws on the difficulty of providing a strict demonstration for doctrines which the philosophers adopt as a matter of course. So it is with the existence of the causal nexus. This can never be demonstrated, for we only observe

---

41 Thus S. Pines in the 1936 classic *Beiträge zur Islamischen Atomenlehre* (repr. New York 1987), also in English as *Studies in Islamic Atomism*. See M. Fakhry, *Islamic Occasionalism and its Critique by Averroes and St. Thomas Aquinas* (London 1958); H.A. Wolfson, *The Philosophy of the Kalam* (Cambridge, MA 1976), 466–600; A. Dhanani, *The Physical Theory of the Kalam* (Leiden 1994); and R.M. Frank, 'Ash^{c}arite Ontology: I Primary Entities,' *Arabic Sciences and Philosophy* 9 (1999), 163–231.
42 Richard Frank has argued for a revisionist view where al-Ghazâlî would actually embrace secondary causes and a largely Avicennian metaphysical scheme, appearances notwithstanding. See his *Creation and the Cosmic System: Al-Ghazâlî and Avicenna* (Heidelberg 1992) and *Al-Ghazâlî and the Ash^{c}arite School* (Durham 1994). In favour of the traditional Ash^{c}arite interpretation see M.E. Marmura, 'Ghazâlîan Causes and Intermediaries,' *Journal of the American Oriental Society* 115 (1995), 89–100 (review of Frank, Creation); 'Al-Ghazâlî's Chapter on Divine Power in the Iqtisâd,' *Arabic Sciences and Philosophy* 4 (1994), 279–315; and 'Al-Ghazâlî on Bodily Resurrection and Causality in the Tahâfut and the Iqtisâd,' *Aligarh Journal of Islamic Thought* 2 (1989), 46–75.
43 TF, 173.11–175.11 and 175.12–178.8.

things happening 'with' (ma^ca) or 'proximate to' (^cinda) other things, never 'by' (bi-) or 'because of' (li-$sabab$) them.⁴⁴ There are simultaneous and subsequent occurrences: but no events can ever be shown to exhibit logical ties with others, and therefore there is no reason to hold that one created thing is ever the cause of the other. Rather, God is the sole agent of everything, instating every occurrence as His will dictates. The pivotal role of Hume's critique of the causal principle is well recognised; but few know that Hume's piercing analysis is predated by more than six centuries by al-Ghazâlî.

The idea that created beings do not initiate their actions but only acquire them (*kasb*) precedes al-Ghazâlî. It was formulated as a solution to early theological controversies regarding man's responsibilities for his actions in the face of God's omnipotence. Accordingly, already Avicenna felt the need to defend the reality of the causal nexus. Avicenna suggested that there is a 'hidden syllogism' at work in the mind's observation that some earthly particular causes something to happen. Every time this happens, what is really observed is only concomitance; so when the mind draws the causal inference, there must be something else at work. According to Avicenna, what happens is that the mind adds to what it sees the proposition that this is what happens either always or for the most part in comparable circumstances. This is the 'hidden syllogism'. It discloses the existence of a universal law working behind the isolated instance.⁴⁵

Alas, Avicenna's explanations only manage to aid al-Ghazâlî in his defence of the occasionalist hypothesis. For al-Ghazâlî can on the basis of Avicenna's exposition claim that the purported causal inference is something which the *mind* puts in things. But it does not do this on its own or by making inferences from nature, no; here as elsewhere the real causal agent is God, who implants in us these beliefs. He also inserts in us the necessary trust in them. God, who has decreed all, has promised that 'we will not find in God's custom any change'⁴⁶ This and not any inherent property in things is the real source of certainty and regularity in nature. 'Demonstration' in the natural sciences is a legitimate practice, but only on account of God's good grace, not by any intrinsic necessity.⁴⁷

---

44 TF, 172.18.
45 Cf., *Ilâhiyât*, 1.1; 1:8.8–10 Madhkûr. Likewise Avicenna, *Shifâ': Kitâb al-burhân* ('Book of Demonstration'), ed. A.E. Affifî, rev. I. Madkûr (Cairo 1956), 249–50. Part of al-Ghazâlî's criticism seems to have been presaged by Ibn al-Tayyib al-Bâqillânî (d. 1010), so Avicenna may have been directing his words against him: see al-Bâqillânî, *Kitâb al-tamhîd*, ed. R.J. McCarthy (Beirut 1957), 43–44.
46 Q. 33:62; 35:42; and 48:23.
47 Cf. TF, 174.16–175.11. On al-Ghazâlî's occasionalist interpretation of demonstration see M.E. Marmura, 'Al-Ghazâlî and Demonstrative Science,' *Journal of the History of Philosophy* 3 (1965), 183–204 and by the same author, 'Al-Ghazâlî's Attitude to the Secular Sciences and Logic,' in: G.F. Hourani (ed.), *Essays on Islamic Philosophy and Science* (Albany, NY 1975), 100–11. One could remark that throughout this exchange, Avicenna wanted to play Kant, while

A brief comment on the theoretical significance of al-Ghazâlî's statements. Although they are mainly geared towards proving the possibility of miracles (a comparatively modest aim), al-Ghazâlî's arguments betray a thought-out systematic view. Al-Ghazâlî accepts that the rules of logic delimit what can take place at one single moment of time. Thus, the limits of logic define the limits of reality, and these are limits not even God can cross. At the same time, he stresses that logic does not dictate what happens between consecutive states. Regularities in nature observable over the course of time are there only by the decree of God, and they could in principle be ordained otherwise. This resembles the Latin medieval distinction between *potentia Dei absoluta* and *potentia Dei ordinata*.[48]

Now, it has often been said that the Latin view allowed for a distinction between logical possibilities and natural possibilities, the latter becoming a subset of the former. The relative 'contingency' of natural laws, in turn, would have led the medievals to question former conceptions of laws of nature and whether they could have been other than what they were generally thought to be. Bearing this in mind, it is extremely interesting to read what al-Ghazâlî has to say about miracles in the context of his second causal theory in the *Incoherence of the philosophers*. The working premise here is that developments in the world are the result of natural potencies being actualised. The question is whether this leaves open the possibility of God working in mysterious ways. Al-Ghazâlî maintains that

> The denial of this is only due to our lack of capacity to understand, [our lack of] familiarity with exalted beings, and our unawareness of the secrets God, praised be He, [has installed] in creation and nature. Whoever studies inductively [*istaqra'a*] the wonders of the sciences will not deem remote from the power of God, in any manner whatsoever, what has been related of the miracles of the prophets.[49]

This is due not to the absolute power of God but rather to the fact that 'the principles of dispositions are beyond our enumeration, the depth of their nature beyond our ken' (178.2–3): thus, the case al-Ghazâlî argues is not one of sheer irrationalism. Rather, al-Ghazâlî recommends an open-minded inquisitive ap-

---

al-Ghazâlî insisted on reading him as Hume: where Avicenna saw the mind's 'hidden syllogism' as disclosing an objectively necessary feature in the way nature outfolds in front of us, al-Ghazâlî interpreted this as a merely psychological mechanism.
48 See TF, 179.5–7. For al-Ghazâlî's modal theory and its significance see the author's 'Possible Worlds in the Tahâfut al-falâsifa. Al-Ghazâlî on Creation and Contingency,' *Journal of the History of Philosophy* 38/4 (2000), 479–502. On the conceptual background see S. Knuuttila, *Modalities in Medieval Philosophy* (London 1993).
49 TF, 178.5–8.

proach. Posited causal rules should be treated as inductions rather than deductions, as generalisations of what has been known to happen rather than metaphysical prescriptions about what can happen.

Averroes' refutation of occasionalism has been expertly recounted in several studies. Let me only briefly outline the main points. According to Averroes, al-Ghazâlî's outlook is epistemologically defeatist. It is contrary to the very nature of man 'in so far as he is man'. If man is a rational animal, yet there is no rationality to be discerned in nature, then all of man's efforts to understand nature are futile and in vain. Knowledge, after all, is knowledge of causes. Thus when the theologians claim that our 'knowledge' is 'certain' because God directly implants it in us, all they manage to do is to make a mockery out of what is meant by both knowledge and humanity.[50]

Nor is this nihilism limited to the epistemological side of things. The denial of secondary causes results in the dissolution of the natural world itself. A denial of natural causes implies a denial of true/essential natures, since it is by their actions that beings are distinguished from each other. Without natural actions we are thus left with a world where 'all is one – indeed, not even one'. For oneness (unity) itself is achieved only by a kind of consistency, by subsistence, and this is precisely what the theologians deny when they deny the permanence (*bâqî*) of attributes.[51]

Averroes' understanding of the necessity of the causal nexus emphasises the intrinsic value of God's creation. If created things do not have a measure of autonomy, they are hardly worthy manifestations of God's creative capabilities. A related facet of creation is its hierarchical organisation. Averroes contends that the Holy Qur'ân itself attests to this hierarchy when it recounts the various ways in which the human being is constructed:[52]

---

50 TT, 521–25 (based on Aristotle, *Metaphysics* 1–2); cf. also TT, 417 and 479; Tafsîr, 9, comm. 7 (2:1135 Bouyges, corresponding to AOAC, 8:f.231H); and *Kashf,* 166 Jâbirî. For studies see first the magnificent Kogan, *Averroes and the Metaphysics of Causation,* 71–164; then A.L. Ivry, 'Averroes on Causation,' in: S. Stein & R. Loewe (eds.), *Studies in Jewish Religious and Intellectual History* (University, Alabama 1979), 143–56; also Fakhry, *Islamic Occasionalism,* 87ff.
51 TT, 519–21. For the problems associated with permanence, see Frank 'Ashᶜarite Ontology: I,' 195ff. and Averroes, TT, 136–41. A similar criticism is made by Moses Maimonides in the *Dalâlât hâ'irîn* ('Guide of the Perplexed'), 1.76. Averroes argues against the occasionalists extensively also in the *Kashf,* 166–71 Jâbirî, and in his commentary to book 9 of the *Metaphysics.* For the latter a critical edition of the Latin translation is available in addition to the Bouyges Arabic: see B. Bürke, *Das Neunte Buch (Theta) des Lateinischen grossen Metaphysikkommentars von Averroes* (Bern 1969).
52 23:12–14, trans. Dawood.

> We first created man from an essence of clay;
> then placed him, a living germ, in a safe enclosure.
> The germ we made a clot of blood,
> and the clot a lump of flesh;
> this we fashioned into bones,
> then clothed the bones with flesh,
> thus bringing forth another creation.
> Blessed be God, the noblest of creators!

For Averroes, the scriptural passage decisively refutes the atomists' notion of simple atoms being attached to (equally simple) accidents. In truth, reality comes laid out in several layers of complexity, all of which can be differentiated in a meaningful manner. The four elements intermix and thereby bring forth composite bodies which in turn serve as material for the emergence of animated and, ultimately, rational beings.[53] Each time, what is brought forward is 'another creation': a new level of form and of 'information', a new functional entity with new powers of action and passion. The interplay of potency and actuality provides the key to this rise in complexity. Different beings act on each other in turn as agents and patients, each bringing forth the particular potencies proper to each.

The theologians were unable to appreciate this because of their inability to understand the process nature of reality. 'In their view, existents only possess two states, that of simple nonexistence and that of actual existence': thus the theologians 'perceived the agent in the way the weak of sight perceive the shadow of a thing instead of the thing itself. And they supposed this to be the thing [itself].'[54] Which is to say: the only way the theologians understood causation was as the origination of things, and, on the strength of this reduction, they inferred that God causes, ie. creates everything: literally *every* thing. This is at best a cookie-cutter version of causality, though, a monochrome silhouette in comparison with the full spectrum of natural causes and the sliding scale of potentiality and actuality. In presenting their model, the Ash^carites diminished the true wonders of creative agency, that of God as well as that of nature.

6. *Sufficient Reason: Arbitration of Inclination*

Averroes sees the order of creation as being determined to a high degree. This connects with his high (and highly idiosyncratic) view of the divine Wisdom, which he sets against al-Ghazâlî's notion of the divine Will in the opening discus-

---

53 TT, 540–41.
54 TT, 132.16–133.1, 133.8–9.

sion of the *Tahâfut* debate. Let us examine this contrast, for it serves to bridge the concepts of causality and determinism – the third in our series of three.

The question is whether God could have had any specific reason to create the world at a given time compared to other possible times, presuming that the world was created in time. Al-Ghazâlî presents the philosophers as arguing that He could not.[55] His counterargument is ingenious. He maintains that it is precisely the function of the will to differentiate between cases in which no ulterior motives can be found. The will is constructed with the aim of distinguishing between similars. The example he adduces bears a striking resemblance to the case of Buridan's ass. If a hungry man were given two dates but no means of deciding between them, then he would die of hunger sooner than take either one of them. But this is evidently absurd: so there must be a means to make the decision. This is where the faculty of will comes in; its instantiation in the natural world is an imitation of the original divine Will.[56]

The introduction of the divine Will as the supreme determinant (*murajjih*) proves most useful for al-Ghazâlî, for it can be invoked wherever signs of contingency and apparent arbitrariness appear in the cosmic order. Al-Ghazâlî sees the world order as contingent to some degree and God's power (*qudra*) as the agent that realises that world to which Will gives preponderance.[57] What is more, al-Ghazâlî is of the opinion that such arbitrary choices cannot be avoided even on the philosophers' own assumptions. As examples al-Ghazâlî cites the placement of the Earth's poles and the overall direction of the heavenly rotations. No earthly reason can be found for why these are determined exactly as they are.[58] This assertion, however, Averroes vigorously denies. It is a sad testimony to the limit of our epistemic capabilities that 'many things which by demonstration can be seen to be necessary seem at first sight merely possible', i.e., contingent.[59] But this fact should not detract from the reality of the situation: for Aristotle had pondered the number and order of the celestial movements and concluded that

> if a single one of these movements should stop, it would result in the corruption of this proportion and order. For it is evident that... the number of these movements is [what it is] either because it is necessary that it be so, or because it is for the best.[60]

---

55 This is a variation on the 'why not sooner?' argument from God's immutability, for whose history see Sorabji, *Time,* 232ff.
56 TF, 22ff.
57 Cf., e.g., TF, 40.7–15.
58 TF, 25.6–14, 26.17–19.
59 TT, 44.13–14.
60 TT, 47.3–7.

In Averroes' view, the universe is arranged in such an optimal manner that not a single cog in the celestial or sublunary clockworks could be taken out or put in without disrupting the whole.[61] As proof Averroes refers the reader to Aristotle's demonstrative works, presumably the *Metaphysics*. Unfortunately, and much to Averroes' embarrassment, Aristotle's remarks in that work are highly speculative, mostly inconclusive, and so opaque as to be offputting.[62] But, no matter: Averroes concludes that it is the principle that counts, and this is stated clearly enough at the closing of *Metaphysics* Lambda. 'The rule of many is not good; let there be one ruler' (1076a3-4). For the world order to be good, it must be traceable back to a single First Principle, which is the mind of God. The content of the divine mind is immutably the same, and nature strives to imitate it: so the divine wisdom fully determines the form the world takes to the exclusion of all contingency.

We can see from the example of the heavenly rotations that al-Ghazâlî and Averroes work with a very different conception of the common Principle of Sufficient Reason. For al-Ghazâlî, this principle primarily indicates the presence of a Will: a free causal agent intervening in every situation where nature does not suffice and reason cannot be found. For Averroes, the same principle indicates that nature *always* suffices: there is always a reasonable cause for every occurrence. Averroes presents the point succinctly when he discusses the role of differentiation (*takhsûs*) in the two parties' metaphysics:

> The differentiation which the philosophers infer is not the one the Ash^{c}arites mean. For the Ash^{c}arites wish [to understand] by differentiation that whereupon a thing is distinguished from something similar or from its opposite, without there being any wisdom in the thing itself to judge between one of the two opposed [things]. The philosophers, as it happens, want [to understand] by differentiation that which is judged on the basis of the wisdom of the product, namely the final cause. For there is not according to the philosophers a single quantity or quality in existence which would not have a final cause based on wisdom...[63]

Averroes goes on to remark that in denying that every existing thing in the world has a sufficient reason in the sense of a final cause and an end the theologians

---

61 TT, 129.
62 For these see *Met.* 12.8 and Averroes' comments *ad loc*. In the *Tahâfut al-tahâfut* Averroes is led to confess rather demurely that the number of celestial movements totals 'somewhat over 40', the exact number being a topic of discussion (apparently it is either 47 or 55): TT, 192.5.
63 TT, 412.2-7. On the corresponding two understandings of 'determinant' (murajjih) in Islamic thought see W.L. Craig, *The Cosmological Argument from Plato to Leibniz* (London 1980), 54ff.

'had denied the reality of the First Creator's wisdom and deprived Him of His very noblest attribute.'[64]

This tells us something important. Of the three basic attributes of God (omnipotence, omniscience, essential goodness), Averroes considers wisdom to be the most fundamental. The divine Wisdom is omnipresent and, as it can be only one, it determines the way reality unfolds in a conclusive manner. Among the Ash^carite theologians, by contrast, the weight of opinion was in favour of giving preponderance to omnipotence. According to the Ash^carites, the divine Might is sufficient reason for the world to be. The rest of the world only gains ontological status as possible objects to God's power [^{c}*alâ qudrati-hi*]. Avicenna seems to favour the third possible option. For him it is God's essential goodness that determines the outward form of the world. God realises and brings to existence all He can, for it is in His nature to do so and a natural outcome of His infinite power to 'overflow'.[65] None of three emphases is likely to have been meant as denying the importance of the other two attributes; and all three certainly have pious visions backing them. But all of them appear to steer dangerously close to determinist notions, albeit in different ways. Let us briefly look at these.

7. *Necessity and Determinism*

As Lenn Goodman once noted, early Muslim theologians found endless reason to call each other 'determinists'. This form of disputation was developed almost to an art. After all, a libertarian with respect to God is a determinist with respect to man, and *vice versa*.[66] When the philosophers arrived on the scene, the question was seen to be yet more complex. There was already a wide literature on the subject focusing for the most part on the famous 'tomorrow's sea battle' of Aristotle's *De interpretatione*. Does the fact, or knowledge of the fact, entail its necessity or even 'cause' it in some way? Are there future contingents?

Concerning the relation between knowledge and necessity, most Muslim thinkers followed the first major Arabic logician, Abû Nasr al-Fârâbî (c. 870–950), 'the second teacher'. In his commentary on *De interpretatione,* al-Fârâbî remarks that the necessity of a certain state of affairs is different from the necessity with which a true statement corresponds to such a state. Thus, knowledge that tomorrow Zayd will leave for a journey necessarily entails that this is what will happen; but it does not entail that Zayd's journey should be 'intrinsically necessary, or that his power [*qudra*] to refrain from travelling should be eliminated. The fact is that Zayd will

---

64 TT, 413.14–15.
65 *Ilâhiyât,* 6.5 and 9.1; 2:296–98 and 2:373–81 Madkûr.
66 Goodman, *Avicenna,* 120.

have the possibility of staying at home.'[67] Causal powers are set apart from mere spectatorship here: the latter does not itself effect the existence of anything. Thus divine foreknowledge does not in itself entail determinism.[68]

Interestingly, al-Fârâbî also mentions the opinion of those who say that what is eternal could nevertheless be otherwise: that the eternal world could just as well not be, or that it remains within God's powers to have been unjust although He never actually is so. This goes to further show that truth, even eternal truth, does not imply determinism.[69] Al-Fârâbî himself does not believe that eternal states of affairs have possibilities for contraries, however, which presents a problem. If eternal states of affairs are considered logically necessary, then, with an eternal world, we seem to have two equally necessary entities: God and the world. The theologian al-Bâqillânî (d. 1010) promptly put forward the following puzzle: if we have two co-eternal principles, each of which logically entails the other's existence, how do we determine which is the cause and which is the effect?[70] In other words, does not the world in this case determine the conditions of God's existence, just as God determines the world's existence?

Avicenna picked up the question in the *Metaphysics* of his *Shifâ'* in a chapter entitled 'On the Prior, on the Posterior, and on Origination'. According to Avicenna, we must distinguish between cases in which the prior condition is necessary for the existence of the posterior but does not necessitate it and the cases in which the prior does necessitate the posterior.[71] This would correspond to the distinction between the necessary and the sufficient conditions of an event. An effect

---

67 Abû Nasr al-Fârâbî, *Sharh li-kitâb Aristûtâlis al-ᶜibâra*, ed. W. Kutsch & S. Marrow (Beirut 1960), 99.11–13. English translation in *Al-Farabi's Commentary and Short treatise on Aristotle's De Interpretatione*, trans. F.W. Zimmermann (Oxford 1981), pagination to the Arabic indicated.
68 Cf., further M.E. Marmura, 'Divine Omniscience and Future Contingents in Alfarabi and Avicenna,' in: T. Rudavsky (ed.), *Divine Omniscience and Divine Omnipotency in Medieval Philosophy. Islamic, Jewish and Christian Perspectives* (Dordrecht 1985), 81–94. It seems that these kinds of questions were current in the Baghdadi philosophical circle: see C. Ehrig-Eggert, 'Yahyâ ibn Adî: Über den Nachweis der Natur des Möglichen. Edition und Einleitung,' *Zeitschrift für Geschichte der Arabisch-Islamischen Wissenschaften* 5 (1989), 284–97; Arabic section 63–97, at 66–72. Already Plotinus had pointed out that 'we must necessarily think of being as preceding thought' (*Enneads* 5.9.8.11–12).
69 *Sharh*, 99.29–100.13 Kutsch & Marrow. Plato is mentioned in the text as saying that the eternal could not be: Zimmermann believes this should in fact read 'Philo' (Al-Farabi's Commentary, 95, n. 1), but cf. n. 52 below. 'Asîdûs', who is still more difficult to identify, is mentioned as taking the latter stance (that God could be unjust). On God's possible wrongdoing see J. van Ess, 'Wrongdoing and Divine Omnipotence in the Theology of Abû Ishâq an-Nazzâm' and R.M. Frank, 'Can God Do What Is Wrong?' Both in: Rudavsky (ed.), *Divine Omniscience*, 53–79.
70 For this challenge, see al-Bâqillânî, *Kitâb al-tamhîd;* 48 McCarthy.
71 Avicenna, *Ilahiyât*, 4.1; 1:164–65 Madkûr. The chapter (pp. 163–69) together with valuable remarks is translated by M.E. Marmura in 'Avicenna on Causal Priority,' in: Morewedge (ed.), *Islamic Philosophy and Mysticism*, 65–83.

may necessarily have *a* cause, but there are typically several different causes that could do the job. By contrast, a fully defined cause only has a single determinate effect. The causal asymmetry provides a further irreducible dimension to the logical relation, which might otherwise be thought to be mutually implicative either way.

Avicenna's example is of the key and Zayd, who opens the door by moving the key with his hand. It seems that neither can happen without the other: yet Avicenna claims that our mind 'is not repelled at all by our saying, "when Zayd moved his hand, the key moved", or, "Zayd moved his hand, then the key moved", but is repelled by our saying, "when the key moved, Zayd moved his hand".' Not that there is a special grammatical or mental rule stating that certain expressions are allowed while others are not; rather, the mind correctly perceives an ontological and causal relation which yields primacy to one event over the other.[72] The door may be opened by Zayd or by ᶜAmr: but if Zayd performs this precise action, it cannot have any other primary effect. (Avicenna had not heard of chaos theory.)

This allows Avicenna to argue that even an eternally and necessarily emanated world is still dependent in a relevant sense on God. The contingency of the eternal was, as is known, central to Avicenna's modal metaphysics. Even the eternal world is in itself always merely possible of existence: this leaves it constantly contingent upon God. And becoming necessary *through* God's actions only compounds this contingency.[73] But Avicenna's scheme is left open to other charges. If generosity is essential to God's nature and if the world is the best of all possible worlds, is God then an automaton who cannot help but give maximally of Himself? Ironically, this question was raised against al-Ghazâlî, since the latter had, in a turn of phrase echoing Neoplatonic sentiments, intimated that 'there is not in possibility anything more wonderful than what is'.[74]

In our view, the problem faced by al-Ghazâlî is somewhat different. For in al-Ghazâlî's occasionalist scheme, God as the sole causal agent is alone responsi-

---

72 Avicenna, *Ilahiyât*, 4.1; 1:165 Madkûr. An close parallel is quoted by al-Ghazâlî in TF, 30–31. It seems possible that Avicenna's thought develops ancient Neoplatonic themes: compare the above with S.E. Gersh, *Kinêsis Akinêtos: A Study of Spiritual Motion in the Philosophy of Proclus* (Leiden 1973), 11–15 and ff.
73 For accounts of Avicenna's argument see M.E. Marmura, 'Avicenna's Proof from Contingency for God's Existence in the Metaphysics of the Shifâ', *Mediaeval Studies* 42 (1980), 337–52; Davidson, *Proofs*, 281–304; and Goodman, *Avicenna*, 61ff. It seems that Avicenna's proof from contingency has more than a little to do with Plato's *Timaeus* 41b–c and its interpretative history, which is why I do not necessarily agree with Zimmermann altering al-Fârâbî's text when the latter says that Plato thought that the eternal could not be (cp. n. 48 above). See my 'Infinite Power and Plenitude: Two Traditions on the Necessity of the Eternal,' in: J. Inglis (ed.), *Medieval Philosophy and the Classical Tradition in Islam, Judaism, and Christianity* (Richmond 2001).
74 On the theological debate that ensued, see E.L. Ormsby, *Theodicy in Islamic Thought: The Dispute over al-Ghazâlî's 'Best of All Possible Worlds'* (Princeton 1984).

ble for every created thing and every conceivable occurrence. Regarding God's 'arbitration' (*hakâm*), then,

> The religious profit to be gained from beholding this attribute of God is to know that the matter is settled and not be appealed. For the pen is already dry, [having written] what exists. The causes are already applied to their effects, and their being impelled towards their effects in their proper and appointed times is a necessary inevitability. Whatever enters into existence enters into it by necessity. For it is necessary that it exist: if it is not necessary in itself, it will be necessary by the eternal decree which is irresistible. So man learns that what is decreed exists, and that anxiety is superfluous. As a result he will act well in seeking his livelihood, with a tranquil spirit, a calm soul, and a heart free from disruption.[75]

The middle section clearly echoes Avicenna, but the end of the passage raises independent problems which al-Ghazâlî is helpful enough to point out. First, what is the relevance of the moral exhortation not to worry when worry, too, comes from God? Second, 'why work, when the cause for happiness or distress has already been settled?' Al-Ghazâlî is reduced to drawing on the powers of paradox at this point. 'For a similar case', al-Ghazâlî suggests,

> take someone who wants to be a jurisprudent and reach the rank of *imâm*. If it is said to him: work hard, learn, and preservere [*sic*]! he will say: 'If God – great and glorious – decreed from eternity for me to be an *imâm*, then no effort will be needed; yet if he decreed for me to be ignorant, then no effort is required.' He should be told: 'If He gave this thought power over you, it shows that He has decreed for you to be ignorant'.[76]

This is certainly a snappy answer; but it still leaves us with a logical puzzle. How do we acquire responsibility for our actions if 'the pen is already dry'? Here God is emphatically not a mere spectator; instead, recording our actions is in His case functionally equivalent to decreeing them.

A similar problem, albeit with a different emphasis, applies to the philosophers' Neoplatonically tinged theory, which also attaches a causal function to God's knowledge. Thus, we may ask Averroes whether he believes that the present world order is conceptually necessary; and he would probably answer in the affirmative. If the divine Wisdom is considered to be isomorphic with creation, then there is really no way in which the world could have come out otherwise than it

---

75 Al-Ghazâlî, *Al-maqsad al-asnâ*, 103.1–6 Shehadi; translation taken from Burrell and Daher.
76 Al-Ghazâlî, *Al-maqsad al-asnâ*, 103.8–11 and 104.8–12 Shehadi; tr. Burrell and Daher.

did. The main differences are that in Averroes' theory, worldly particulars have autonomous causal powers which they exercise; and that in al-Ghazâlî's theory, God, even in decreeing this world to come about, still retains the choice between several different natural orders of equal cardinality. This leads us to a final observation.

8. *Conclusion*

It has sometimes been remarked that the monotheist notion of a willed creation *ex nihilo* tends to produce systematically different results from an approach in which the present world order is simply taken as a given, as eternal, or as necessarily what it is. In this paper I have tried to illustrate what this could mean. As should be expected, the results are more than a little ambivalent. If history has lessons to teach us, then these are rarely neat or particularly clear-cut. But on the basis of the Islamic debate one rather general conclusion may be tentatively voiced. It has to do with the question of how 'transcendent' or 'immanent' we take God to be in respect to creation.

For al-Ghazâlî, it is crucial that God be an utterly transcendent and free agent. He seems to believe that for this to make sense God must retain a certain autonomy with respect to the world He has created. This involves a relevant choice between alternative worlds. The resulting picture lends the created world a certain precariousness, but, on the other hand, renders it open to several different interpretations, a possible incentive for a 'possible worlds' approach to nature. If, on the other hand, God is regarded as somehow participating in the world, then it would seem that all of the world's features reflect the divine wisdom. This, we have seen, is how Averroes views the situation.

What we win in the latter case is the conviction that all the parts of our theory will ultimately 'add up'. What we lose is the perspective that they could be fundamentally other than what we have formerly taken them to be. One approach, then, opens to infinity: the other forecloses this option in favour of a finitely knowable universe. What is ironic is that it is the latter notion which can claim to countenance at least one kind of infinity (that of infinite time) within its worldview; the former strictly binds the world to a finite framework. Perhaps this is an aspect of the interplay between mathematical and metaphysical interpretations of infinity.

3. God, Causality, and Nature

Some Problems of Causality in Medieval Theology

Reijo Työrinoja (Helsinki)

1. Introduction

Modern technology has made it possible to harness almost all of nature to serve human needs. But how has this idea come into being historically? It is obvious that technological interest presupposes a certain kind of view of nature and the laws directing it. In what follows I will discuss certain important conceptual changes considering the notions of cause and effect both in the Middle Ages and at the beginning of the Modern Times. My intent is to show how some of those new ideas made possible the idea of using and manipulating material nature. To do that I will describe how these conceptual changes prompted technological interest. I will consider first Thomas Aquinas's view of causality, and thereafter how William Ockham revised the issue. Finally I will comment on Luther's and Calvin's views on causality.

2. *Aquinas on the Divine Causality*

As a theologian, Thomas Aquinas in his view of causality[1] states that God has by one unlimited act created the prime matter (*materia prima*) with the substantial forms which constitute the natural species, and the secondary causes which direct the natural activities of beings.[2] By virtue of creation, each being has its own ac-

---

1 For Aquinas's view of causality, his commentary on the Pseudo-Aristotelian work *The Book of Causes* (*Liber de expositione bonitatis purae seu Liber de causis*) has great importance. Gerard of Cremona translated it from Arabic to Latin in 1187. The work is a paraphrase of Proclus's *Elements of Theology* (Elementatio theologica) written by an unknown Muslim author, a matter of which Thomas was conscious. *In librum de causis. Expositio,* Cura et studio fr. Ceslai Pera, O.P. cum introductione historica Sac. Petri Caramello et praeludio doctrinali Prof. Caroli Mazzantini (Taurini 1955) (hereafter *In librum de causis*), Proeem. 4–5;9. Aquinas's socius William of Moerbeke translated Proclus's book in 1268. Aquinas refers repeatedly in his other works to the *de Causis,* especially in the disputation on the Truth. See *Quaestiones disputatae. Volumen I. De Veritate,* Cura et studio O. Fr. Raymundi Spiazzi (Taurini 1964) (hereafter *Quaestiones disputatae*), q.5, a.9, 10.arg.; q.6, a.8, 9.s.c.; q.5, a.9, 10.arg.; q.6, a.8, 9.s.c.; q.24, a.14,c; *Summa theologiae. Pars prima et prima secundae* (Taurini 1952) (hereafter Summa theologiae), Ia, q.14, a.2, ad1.
2 *In librum de causis,* Pr.XXIV, lec.24, 125. – 388; Pr.XX, lec.20, 121;399. – See also Jan Aertsen, *Nature and Creature: Thomas Aquinas's Way of Thought* (Leiden 1988), 92–140, 279–336.

45

tive power by which it can, together with the divine primary causality, act as an immediate cause of physical effects.[3] For Aquinas, every being in the corporeal world is a composition of matter and form. On the metaphysical level, reality is already divided into classes of beings so that each particular being is a representative of some species, and connected to another representative of the same species by a common substantial form.[4] Each singular being realizes in its life history the essence of its own species. The basis of this individual actualization of a species, and the principle of its individuality, is matter. Even if representatives of a species are identical with each other in relation to the common substantial form, they are numerically and accidentally distinct individuals in relation to matter, which makes them particular beings with different accidental features.[5] The substantial form is the basis of all effects which a particular thing is capable of causing in the world. Thus, for example, the capacity of fire to cause heat is based on the substantial form of fire.[6]

In Aquinas, every singular entity is 'a thing subsisting in itself' (*ens in se subsistens*), even though its existence ultimately depends on God who alone is a true being (*ipsum esse subsistens*) without any other cause than himself.[7] As a creature, every being is part of the divine providential program, and as such, it is capable of producing characteristic effects for aiming at its end and actualizing its own essence.[8] The final end, however, common to all beings, is God himself. God is the formal and final cause of every movement and action but not the immediate efficient cause (*causa immediata*). God, who is the first efficient cause, does not need to intervene in the actual world every moment because the substantial forms and secondary causes take charge of their tasks in relatively independent ways.[9] Hence, by emphasizing the real nature of substantial forms and secondary causes,

---

3 *In librum de causis*, Pr.I, lec.1, 28; 6. A secondary cause, however, cannot be the cause of the whole species (causa speciei). For example, a particular human being cannot be the cause of the human species because it would then also be its own cause, which is impossible. *Liber de veritate Catholicae Fidei contra errores Infidelium, seu Summa contra Gentiles. Vol. III*, Cura et studio Ceslai Pera Petro Marc et Petro Caramello (Taurini 1961) (hereafter *Summa contra gentiles*), III, 65, 2400.
4 *Summa theologiae*, Ia, q.75, a.4, c.
5 *Summa contra gentiles*, III, 65, 2400; *Summa theologiae*, Ia, q.75, a.4, c.
6 *Summa theologiae*, IaIIae, q.2, a.9, ad1.
7 *Summa theologiae*, Ia, q.14, a.2, ad.1; Ia, q.13, a.11, c.
8 *Summa theologiae*, Ia, q.19, a.4, c.
9 *In librum de causis*, Pr.XXIII, lec.23, 120;386. – On the relation between the first and secondary cause, see also Aertsen's discussion. E. Przywara, for example, emphasizes the importance of secondary causes using the expression Grundgesetz der 'causae secundae'. For him, secondary causes assure that creatures have independent being and action that manifests God as the first cause only indirectly. Aertsen, however, criticizes Przywara's interpretation that it gives too much emphasis to the autonomy of created reality. – 'The autonomy of the natural is not a sign of God's insufficiency but rather of His goodness.' Aertsen, *Nature and Creature*, 322–323. *In librum de causis*, Pr.XXIII, lec.23, 120;386.

Aquinas makes room for the relative independence of the created existence as an immediate cause of its acts.[10] It is characteristic of Aquinas's view that in the natural sphere, *in rebus naturalibus,* neither of these metaphysical elements, matter and form, exists as separated, but they exist together in the shape of a concrete, particular being. It follows from this that there is no pure, 'dark' matter which has not the potency to become something. Matter, in this model, is understood as potentially active, naturally capable of receiving different forms which will be actualized as particular beings.[11]

It is plausible to think that this kind of conception of the relation between matter and form is not apt to stimulate any thorough-going technological interest in nature, let alone any manipulative approach. Human artificial action can only imitate original substantial forms and secondary causes based on them. As an imitation, it can never be better than nature itself, which is the paradigmatic model for any action. For conceptual reasons, it would be unnatural for human actions to attempt to surpass nature. Neither can artifacts, produced by human beings, have any real ontological status in the 'great chain of being.' From the metaphysical point of view, compared with God and created natural beings, artifacts are only ontological parasites, on the lowest level of reality, and not truly beings.[12]

For Aquinas, even though God has created the world by his free will, its metaphysical basis is God's visionary knowledge (*scientia visionis*) of his own eternal

---

10 *Summa theologiae,* Ia, q.105, a.3, c. – 'Ad primum ergo dicendum quod operatio intellectualis est quidem ab intellectu in quo est, sicut a causa secunda: sed a Deo sicut a causa prima. Ab ipso enim datur intelligenti quod intelligere possit.' *Summa theologiae,* Ia, q.105, a.3, ad1; Ia, q.105, a.5, c.
11 *Summa theologiae,* Ia, q.66, a.1, c; Ia, q.66, a.1, ad3.
12 Ockham, for example, thinks differently here. While commenting on the second book of Aristotle's Physics, Ockham distinguishes between natural and artificial things. But, what is essential is that he does not base this distinction on the metaphysical interpretation of the natural and the artificial. According to him, the distinction between natural and artificial things is not based on the real definitions of things but on nominal definitions. Nominal definitions (quid nominis), unlike real definitions, concern the use of words in language, and in this sense it is true that the words 'natural' and 'artificial' have different meanings. This does not mean, however, that they were contrary predicates which cannot speak of the same thing at the same time. It is not prima facie contradictory to say that natural things are artificial, or vice versa that artificial things are natural. *Expositio in Libros Physicorum Aristotelis. Opera Philosophica IV* (St. Bonaventure, N.Y. 1985), Lib II, lc,4n.,6. On the distinction between real and nominal definitions, see *Summa logicae. Opera Philosophica I* (St. Bonaventure, N.Y. 1974), Pars III, cap. 28, 556;34–35. According to Ockham, anyone can speak without knowing the real definition of a thing, but not without knowing its nominal definition. *Summa logicae,* OP I, 555;11–556;14. – *Quodlibeta septem. Opera Theologica IX* (St. Bonaventure, N.Y. 1980), Quodlibet V, q.19, OT IX, 554; 16–17. Only the absolute names ('human being', 'lion', 'goat', etc.), which signify things composed by matter and form have real definitions. Nominal definitions are by nature connotative and relative names. Quodlibet V, q.19, OT IX, 554; 22–28.

essence. The vision of the triune God comprises all possible forms with which finite beings can imitate his infinite essence. The scope of God's free choice is his own essence which is the metaphysical basis for all possible creatures. Therefore, the whole being, even though it has its existence as a result of God's free choice, is participating, in some sense, in God's being that alone is being *per essentiam*.[13] Thus, from Aquinas's point of view, the created world is not merely an artifact, but not a direct participant in the divine essence, although he often uses the Neo-Platonic terms *emanatio* and *influxus* when describing the creation. As a created being, the world is produced outside the divine essence, and therefore, modes of divine being and created being are only analogical (*analogia entis*) to each other.[14]

3. *Aquinas and Islamic Occasionalism*

Aquinas was, however, familiar with an alternative way of thinking that was in opposition to his own metaphysical and ontological views. This alternative paradigm came from Arabic philosophy, especially from the Asharites movement, particularly represented by Abû Hâmid al-Ghazâlî (1058–1111).[15] In Ghazâlî's conception, all causes are reduced to God's creative action as the first cause, with the result that no secondary causes exist. There are only phenomena which temporally succeed each other, but so that every instant of time is ontologically separated from another instant of time. For example, there is no necessary connection between fire and heat. One can say that the cloth will burn *when* it comes physically into contact with fire, but never that it burns *because* it has come into contact with fire. God is the only cause and active agent in reality, everything else is merely momentary and occasional occurrences, hence the reason why this view is called 'occasionalism.'[16] The model suggests an idea of the continuous creation, that is to

---

13　On Aquinas's view of participation, see Rudi A. Te Velde, *Participation and Substantiality in Thomas Aquinas* (Leiden 1995) and John F. Wippel, 'Thomas Aquinas and Participation,' in: John F. Wippel (ed.), *Studies in Medieval Philosophy* (Washington, D.C. 1987), 117–158.

14　*Summa theologiae,* q.104, a1.c; Ia, q.104, a.1, c. God cannot communicate his capacity for conserving and creating to any creature, that is to say, as the creator he cannot make a creature that is totally autonomous and self-sufficient. This would presuppose that God is not the cause of a creature, which is impossible. *Summa theologiae,* Ia, q.104, a.1, ad2; Ia, q.45, a.5, ad3.

15　On Ghazâlî's life and work, see Iysa A. Bello, *The Medieval Islamic Controversy between Philosophy and Orthodoxy. Ijmâ< and Ta>wî in the Conflict between Ghazâlî and ibn Rush* (Leiden 1989), 6–10.

16　See Richard Sorabji, *Time, Creation and the Continuum. Theories in Antiquity and the Early Middle Ages* (London 1983), 297–306. Al-Ghazâlî's main philosophical work is *Tahâfut al-Falâsifa* (lat. *Destructio Philosophorum*). Averroës responded to Ghazâlî's critique in his work *Tahâfut al-Tahâfut,* which was translated into Latin in 1328 with the name *Destructio Destructionis*. Averroës criticizes Ghazâlî's theory and quotes him extensively. See Harry A. The Wolfson,

say, God creates at each instant of time a new accident in the place of the previous and destroyed accident, however the new accident is similar to the destroyed one and makes one believe that continuity and identity of a thing in time and place is based on its substance. It follows from this that in Ghazâlî's theory there are no proper substances at all, but all things are merely aggregates of accidental properties held together by the divine creative will.[17] If God were to decide not to will a particular thing, which he no doubt could do, it would be immediately annihilated.[18]

Occasionalism was a rival theory to Aquinas's own Aristotelian approach, according to which material reality already includes natures, substances and forms, which are active foundations and real causes of perceptible changes in the actual world. Aquinas was familiar with this occasionalistic theory, on the one hand, through Maimonides's *Guide of the Perplexed,* where Rabbi Moyses, as Aquinas calls him, challenges the truth of several statements of the Mutakallims, and on the other hand, through Averroës's critique in his commentaries on the *Metaphysics* and the *Physics.*[19] In his commentary on Peter Lombard's *Sentences,* Aquinas examines the question of whether there is in nature any other efficient cause than

---

Philosophy of Kalam (Cambridge, Mass. 1976), 594; William J. Courtenay, 'The Critique on Natural Causality in the Mutakallimum and Nominalism,' *The Harvard Theological Review,* 66 (1973); and William J. Courtenay, *Covenant and Causality in Medieval Thought: Studies in Philosophy, Theology and Economic Practice* (London 1984), V, pp. 78–79. – Recently two comparative studies on al-Ghazâlî have been published, one with Augustine, and another with Descartes. See Johan Bouman, *Die Theologie Ghazâlî's und Augustins im Vergleich* (Giessen 1990), and Mahmud H. Zaqzuq, *Al-Ghazâlî's Philosophie im Vergleich mit Descartes* (Frankfurt a.M. 1992). – Descartes's view of final causality is problematic. On the one hand, he seems to eliminate the idea of final causes influencing nature, but some of his formulations are quite similar to the traditional Aristotelian-Thomistic conception. See Peter K. Machamer, 'Causality and Explanation in Descartes,' in: Peter K. Machamer and Robert G. Turnbull (eds.), *Motion and Time, Space and Matter: Interrelations in the History of Philosophy and Science* (Ohio 1976), 177–181.

17 Majid Fakhry, *Islamic Occasionalism and its Critique by Averroës and Aquinas* (London 1958), 56–82 and Sorabji, *Time, Creation and the Continuum,* 300. As fas as I know there is only one study explicitly concerning al-Ghazâlî's view of causality. See Carol Bargeron, *The Concept of Causality in Abû Hâmid Muhammad al-Ghazâlî's Tahâfut al-Falâsifah* (Michigan 1978). Bello, for example, does not deal in detail with this problem in the controversy between al-Ghazâlî and Averroës but touches upon it while discussing the dispute concerning the bodily resurrection. See Bello, *The Medieval Controversy,* 126–141.

18 Also according to Aquinas, all being depends on God's will by which he as the cause of beings sustains and conserves his creation. If God would withdraw his will and conservation, the whole of creation would be annihilated. *Quaestiones disputatae. Vol. II. De potentia,* Cura et studio M.Pession (Romae 1953) (hereafter *De potentia*), q.5, a.3, sed contra, arg. 2. Aquinas believes, however, that such destruction will never happen and that the world with its substances will last forever. *De potentia,* q.5, a.4, c. On Aquinas's theory of annihilation in detail, see James F. Ross, 'Aquinas on Annihilation,' in: Wippel (ed.), *Studies in Medieval Philosophy,* 177–199.

19 *De potentia,* q.3, a.7, c., 56.

God himself. He introduces three different positions. According to the first position, God acts immediately in all things, so that nothing else has causal power over anything (*nihil alius est causa alicuius rei*). Actually, it is not fire that causes heat, but God himself, nor is it a hand that moves, but a hand is directly moved by God. It is not necessary that fire causes heat but this happens only because God so wills.[20] Aquinas's example is parallel to the view criticized by Averroës that the 'Divine One' causes everything directly without any mediating cause (*sine medio*). According to Averroës's critique, the theory suggests that there is no real action in things at all. It follows from this that there are no proper essences either, because different essences differ from one another explicitly by their different acts.[21] Maimonides also says that the Muslim philosophers (*loquentes in lege Maurorum*) have claimed that the relation between heat and fire is only a habit and not the law of nature. Aquinas sees three critical errors in Muslim occasionalism. First, this conception presumes that there are no substantial forms but all forms are accidental.[22] Second, these accidents are merely occurrences or events without any temporal duration, so that no accident lasts two instants (*per duo instantia*). Third, continuity in the sensible world requires that God sustains directly its existence at every instant of time. This means that nothing really exists, but the formation of beings (*formatio rerum*) is a continuous process of becoming (*semper esset in fieri*).[23] God creates in every moment a new similar accident within the same species in place of the destroyed one, which in turn will be destroyed, etc., as long as God wills.[24]

---

20 In Lib.II Sent., dist. I, q.1, a.4; *Summa theologiae*, Ia, q.105, a.5, c. See also *Summa contra gentiles*, III, 65; 88, 2400.
21 Averroës, *IX Metaphys. c. 7* (Venitiis, apud Cominum de Tridino Montis Ferrati 1560), t.VIII, fol. 265a A B. Quoted from the *Summa contra gentiles*, III, cap. 69; 94, 2431, fn. 1.
22 *Summa contra gentiles*, III, cap. 69; 95; 2441.
23 *Summa contra gentiles*, III, 65; 89, 2406.
24 Maimonides describes this view as follows. – 'Vera notio accidentis est quod non durat nec subistit per duo instantia, unde statim atque accidens creatur, desinit nec durat; Deus autem aliud accidens eiusdem speciei creat. Hoc similiter desinit, tertiumque [accidens] eiusdem speciei Deus creat et sic continenter quamdiu illam speciem accidentis Deus durare vult. Si autem Deus in illa substantia, accidentium speciem alteram creare vult, id [quod vult] facit, at si a creando se abstinet nullumque accidens creat amplius, substantia illa esse desinit.' Maimonides, Doctor Perplex, ed. D. I. Maroni. Quoted from the *Summa contra gentiles*, 65;. 89, fn.5. See also Wolfson, *The Philosophy of Kalam*, 589–593.

4. Causality in the Via Moderna

After Pierre Duhem's pioneering work, several scholars have agreed with the importance of the great Parisian condemnation of 1277, declared by the bishop Etienne Tempier, for the later development of scientific ideas in the Middle Ages.[25] Some of these condemned propositions dealt with physical notions of natural philosophy and their theological implications. One of the main targets of the condemnation was a number of deterministic ideas, taught in the Faculty of Arts at Paris, and which called into question divine omnipotence concerning the physical world and its laws.[26] It was forbidden, for example, to teach that God cannot

---

25 A French scholar, Pierre Duhem (d.1916), wrote his celebrated ten-volume *Le Système du Monde* during the years 1909–1916. There he deals in depth with different theories in natural philosophy from Plato to Copernicus. His extensive work can be regarded as a forerunner to the history of the studies of cosmology and physics and of the philosophical and theological ideas behind them. Duhem was the first to show that new scientific ideas are closely connected with certain background theological assumptions and that conceptual changes took place within these ideas in medieval times. He paid special attention to the Parisian condemnations in 1270 and 1277, of which the latter was particularly important. This 'great' condemnation, including 219 propositions, prohibited the teaching of certain extreme, deterministic interpretations of Aristotle at the University of Paris. Duhem emphasized these theological factors to such an extent, however, that his approach may be seen as apologetic. He was particularly criticized for such an approach by another famous scholar in the history of science, Aleksander Koyré. – Pierre Duhem, *Medieval Cosmology: Theories of Infinity, Place, Time, Void, and the Plurality of Worlds* (Chicago 1985), 431–451, includes a translation of parts of Duhem's studies of medieval natural philosophy, such as the discussion on the plurality of worlds. F. Floris Cohen is a Dutch scholar whose recently published study is an extensive and profound research into the discussion on the roots of the scientific revolution. On the critique of Duhem's thesis, see H. Floris Cohen, *The Scientific Revolution: A Historiographical Inquiry* (Chicago 1994), 45–56. Christopher Kaiser, *Creation and the History of Science* (Grand Rapids 1991), gives a more concise survey of the topic. Along with Duhem, Edwin A. Burtt, *The Metaphysical Foundations of Modern Physical Science: The Scientific Thinking of Copernicus, Galileo, Newton and their Contemporaries* (London 1980) represents the older scholarship in this field. See also Reijer Hooykaas, *Religion and the Rise of Modern Science* (London 1972). On the impact of the Parisian condemnation, see David C. Lindberg, *The Beginnings of Western Science: The European Scientific Tradition in Philosophical, Religious, and Institutional Context, 600 B.C. to A.D. 1450* (Chicago 1992), 234–248; Edward Grant, *The Foundations of Modern Science in the Middle Ages: Their Religious, Institutional and Intellectual Contexts* (Cambridge 1996), 70–85. An essential subject of the debate concerns the question of the continuity between medieval and classical physics. Anneliese Maier, unlike Duhem, who strongly emphasized it, denied that there is such a continuity. According to her, there is an opposition between the principle of inertia in classical physics and the medieval Aristotelian principle of omne quod movetur, movetur ab alio. See James A. Weisheipl, O.P., 'The Relationship of Medieval Natural Philosophy to Modern Science: The Contribution of Thomas Aquinas to its Understanding,' *Manuscripta* 20 (1976), 181–196, and André Goddu, *The Physics of William of Ockham* (Leiden 1984), 229–232.

26 *Summa contra gentiles, III, Appendix II: Propositiones damnatae a Stephano Tempier Episcopo Parisiensi. 1277*, 493–502. – Prop. 21, 45, 48.

do anything without secondary causes, that God cannot create a mere accident without a subject in which the accident inheres, or that God cannot create the plurality of worlds.[27] As an after-effect of Tempier's condemnation, theories of contingency in nature became philosophically and theologically more acceptable. The condemnation favored especially the views later developed by John Duns Scotus and William Ockham. According to both of them, God has in his absolute power (*potentia absoluta*) chosen and made the physical laws to direct the common course of nature. This emphasis upon God's free power, restricted only by the law of contradiction, made it natural to view the world from God's perspective, from within his absolute freedom and power. Moreover, this new approach encouraged a new kind of scientific imagination and various thought experiments,[28] for example the idea of the plurality of worlds; an assumption which Aristotle had tried to prove to be impossible and contradictory. To consider the world from the point of view of God's absolute power made the idea of the plurality of worlds comprehensible, even though most medieval writers very likely thought that *de facto* there is one world only. The idea of the plurality of worlds did not mean only that along with our solar system there may exist another earth with its own sun, planets and orbits, similar to ours and without any relation to our world, but

---

27 *Summa contra gentiles, III, Appendix II: Propositiones damnatae a Stephano Tempier Episcopo Parisiensi. 1277,* Prop. 34, 63, 141.
28 As a consequence of the Condemnation of 1277, God's absolute power became a favorable tool for introducing subtle, imaginative questions which generated novel replies. See Edward Grant, 'The effect of the condemnation of 1277,' in: Norman Kretzmann, Anthony Kenny, Jan Pinborg (eds.), *The Cambridge History of Later Medieval Philosophy: From the Rediscovery of Aristotle to the Disintegration of Scholasticism 1100–1600* (Cambridge 1982), 537–539. – 'For the application of these languages (in theology as in natural philosophy) proceeded secundum imaginationem and this, as we have seen, was permitted, even urged, by the invocation of potentia Dei absoluta, a factor in turn at the very center of the rising voluntarism. In other words, since God's will can act with absolute contingency, it follows that, in a given problem, the entities or events involved can exist or can occur all these imaginable ways, where "can" is to be taken de potentia Dei absoluta and where "all these imaginable ways" consist in the alternative casus that are analyzed by our language.' John E. Murdoch, 'Unitary Character of Medieval Learning,' in: John E. Murdoch, and Edith D. Sylla (eds.), *The Cultural Context of Medieval Learning. Proceedings of the First International Congress on Philosophy, Science, and Theology in the Middle Ages – September 1973* (Dordrecht 1975), 292. 'The discovery of the scientific role of imagination allows for mental experiments. Where facts are not in the reach of experience, we grope for the facts with our imagination, the realm of the potentia absoluta, the terra incognita, the unknown realm of logical possibilities.' Heiko Oberman, *The Dawn of the Reformation: Essays in Late Medieval and Early Reformation Thought* (Grand Rapids, Michigan 1986), 195. On the meaning of 'imaginary,' see Edward Grant, *Planets, Stars, and Orbits. The Medieval Cosmos, 1200-1678* (Cambridge 1994), 177–185. On the relation of art, nature, and experiment in medieval Aristotelianism, see William R. Newman, 'Art, Nature, and Experiment among Some Aristotelian Alchemists,' in: Edith Sylla and Michael McVaugh (eds.), *Texts and Contexts in Ancient and Medieval Science. Studies on the Occasion of John E. Murdoch's Seventieth Birthday* (Leiden 1997), 305–317.

also that there might be different worlds with different physical circumstances and laws of nature.[29]

In late medieval nominalism, the question of causality in nature came into a new light. According to Ockham, no valid scientific demonstration can be given for the existence of causality. By 'demonstration' Ockham understands an argument in which the conclusion follows necessarily from necessarily true premises. In the case of causality this would imply that the causal connection is necessary, and not a contingent one, which is Ockham's own position. Although the natural order is contingent, the causal relation is epistemologically justified in that it can be proved inductively by human experience.[30] In the common course of nature, when one has an intuitive cognition that A is present and an intuitive cognition

---

29 'Whether there actually exist other worlds, was not at issue, but rather how the existence of other possible worlds is to be comprehended. – Generally speaking, given the theological assumption that God could have built a different universe, these kinds of arguments seek to explain in Aristotelian terms what would have happened in that case, and which could have been the laws of that different world.' Eugenio Randi, 'Plurality of Worlds: Fourteenth-Century Theological Debates,' in: Simo Knuuttila, Reijo Työrinoja, Sten Ebbesen (eds.), *Knowledge and the Sciences in Medieval Philosophy: Proceedings of the Eighth International Congress of Medieval Philosophy (S.I.E.P.M.) Vol. II* (Helsinki 1990), 322–330, 330. – Few in the Middle Ages held the the view that God could not create other worlds. Peter Abelard and John Wycliff are well-known exceptions to this accepted position. Grant, *Planets, Stars, and Orbits*, 151. From this point of view, the medieval discussion about whether God is able to create a world that is better than our actual one, is particularly interesting. The idea of a world better than the actual world suggests the idea of several alternative worlds, or, in modern terms, the idea of possible worlds. The question can be posed as follows: Did God exhaust all of his possibilities when he created the world? And, if he did not, are there among the non-actualized possibilities worlds that are better than the created world, with better species or individuals? If on the one hand the answer is negative, then God seems to have no free choice in his creation. If the answer is on the other hand affirmative, then it seems that God is not a perfect creator. Ockham sees no problem with the possibility that God could have created a better world had he so desired. God could have created better species and individuals as well and several better worlds. – Ockham deals with this problem in his Commentary on the first book of the Sentences, especially in distinction 44, where the question of whether God can make (facere) a better world (mundum meliorem) is explicitly posed. He also deals with it in distinction 43, question one, where he asks whether God could make something he has not yet made or wished to make. Guillelmi de Ockham, *Scriptum in librum primum Sententiarum. Ordinatio. Distinctiones XIX–XLVIII, Opera Theologica IV*, eds. G. I. Etzkorn & F. E: Kelley (St. Bonaventure, N.Y. 1979), Lib. I, dist. 44, q.1, OTh IV, 655; 9–17. On the discussion of the plurality of worlds, see Maurer, Armand, 'Ockham on the Possibility of a Better World,' in: Maurer Armand, *Being and Knowing: Studies in Thomas Aquinas and Later Medieval Philosophy* (Toronto 1990), 264–266; André Goddu, 'William of Ockham,' in: Kenan B. Osborne, O.F.M (ed.), *The History of Franciscan Theology* (St. Bonaventure, N.Y. 1994), 231–304; Grant, *Planets, Stars, and Orbits*, 150–168; and Grant, *The Foundations of Modern Science in the Middle Ages*, 119–126. On different meanings of 'world' (mundus), see Grant, *The Foundations of Modern Science in the Middle Ages*, 127–131.

30 *Scriptum in librum primum Sententiarum. Ordinatio. Prologus et Distinctio Prima. Opera Theologica I*, eds. Gideon Gál & Stephen Brown (St. Bonaventure, N.Y. 1967), dist. I, q.6. OT I, 497. See Courtenay, *Covenant and Causality in Medieval Thought*, V 79–80, 88–89.

that B is present so that whenever B occurs, A has also occurred before, and B does not occur without A occurring, one is justified in stating that A is the cause of B. But we have no intuitive cognition of causal relation itself, anymore than any other relations. Relations are inferred rationally but not intuitively cognized.[31]

Thus, for Ockham, secondary causes are genuinely explanatory causes in the epistemological sense, even though one can have no ultimate ontological guarantee of the causal connection between two phenomena.[32] For Ockham, as for Aquinas, there are two simultaneous efficient causes of every phenomenon, the natural and the divine. The essential difference between them concerns the ontological status of secondary causes. Ockham rejects the metaphysics of common essences argued by Aquinas, according to whom effects are potentially contained in the substances.[33]

Ockham's position can be described more accurately by comparing it with that of one of his contemporaries, viz. Robert Holkot's view of the same issue. Holkot espoused a more cautious and, in a sense, more skeptical attitude epistemologically to the problem of causality than did Ockham. According to Holkot, one can never know with certainty whether some effect is immediately caused by God without any natural cause, or whether it happens by natural causes without any direct divine causality. Under any circumstances, God can cause everything exactly in the same way as natural causes do; therefore, everything which seems to us to happen by secondary causes may be caused immediately by God.[34] Hence, for Holkot, there can be no rule (*regula*) for demonstrating with certainty (*certi-*

---

31 *Scriptum in librum primum Sententiarum. Ordinatio*, dist. I, q.3. OT I, 418; 10–17.
32 Ockham's view of natural causality is also closely related to his view of theological causality. Both forms of causality have their background in the idea of God's absolute power. By theological causality is represented by following questions: How does the death of Christ cause the salvation of humankind? How do good deeds done in the state of grace cause eternal life? How do sacraments cause grace, etc? Ockham thought that there is not a question of any inherent determinant contained in these theological causes, but their effectiveness is based on divine decree and promise. The sacraments, when they are consecrated in an appropriate way, cause grace ex opere operato, not by any inner cause contained in the elements, however, but by virtue of the covenant and pact instituted by God. On the other hand, in his absolute power God is not constrained by his own decrees and can deviate from them. Ockham thinks, anyway, that de facto God will keep his promise and acts in accordance with the order of salvation prescribed by him, in the same way as he maintains the common course of nature. See Courtenay, *Covenant and Causality in Medieval Thought*, V 91–92.
33 According to Courtenay, Ockham does not take a final position on what this cause ultimately is, but he is satisfied with the definition according to which a cause is whatever being present is sufficient to produce an effect, and being absent is sufficient to prevent it. In this sense Ockham agrees with Aquinas that secondary causes have their own active power, along with God's primary immediate causality. Courtenay, *Covenant and Causality in Medieval Thought*, V 90–91.
34 Robertus Holkot, *Determinationes item quarundam aliarum quaestionum: In quattuor Libros Sententiarum Quaestiones. Ludguni 1518* (Frankfurt 1967), Determinatio, q.III, M.

tudinaliter) that one thing is a cause of another. A can be only a probable cause (*probabiliter*) of B. Causality is based on the spatio-temporal connection between two ontologically singular and separate phenomena. When one repeatedly sees that A's occurence 'in different places and in different points of time' (*in diversis locis et temporibus*) is followed immediately (*statim*) by B. For example, when one sees that when fire occurs, heat also immediately occurs, 'one can say' (*dicitur*) that fire is the cause of heat.[35]

Obviously, an important difference obtains between the positions of Ockham and Holkot on causality. For Ockham, no scientific (i.e. necessary) demonstration for the existence of the causal connection between two phenomena can be given, but to suppose that there is one, is epistemologically justified on the basis of the common course of nature, even if it is deduced from a finite number of contingent cases. From Holkot's point of view, however, no conclusion that something really 'is' the efficient cause of something else is justified. A finite number of occurrences in different places and points of time does not justify any rule in the epistemological sense (*scire*). What it does justify is that one is justified 'to say' (*dicimus*) that A is the cause of B. On the linguistic level, we call something the cause of something else if the former immediately occurs before the latter and the latter does not occur without the former. But to say this, is to say something essential about what the terms 'cause' and 'effect' mean and about their nominal definitions, but one cannot conclude that causality really exists.

5. *Luther and Calvin on Divine Causality*

In the discussion of the question concerning whether God is the immediate cause (*causa immediata*) of everything, Ockham proposes that there is no real distinction between two fundamental divine actions, creation and conservation, because 'nothing positive differentiates them.' Ockham approaches the dilemma from a semantical point of view. He distinguishes between a primary meaning or signification and a secondary meaning or connotation of the terms 'creation' and 'conservation.' The term 'creation' signifies (*significat*) primarily an existing thing and simultaneously signifies secondarily or connotes one and the same thing negatively by denying that any other thing can precede it temporally. The term 'conservation' signifies primarily one and the same existing thing by connoting its temporal continuity and but in a negative sense by denying its interruption or destruction. This means that in both cases the primary signification of the terms 'creation' and 'conservation' is the same but the connotation of both terms is negative. Thus, nothing positive in the primary signification differentiates these terms

---

35 Holkot, *Determinationes item quarundam aliarum quaestionum,* q.III, M.

from one another; the difference is alone connotative and negative. The conclusion is that the terms refer to one and the same divine act and not to two different acts.[36]

The distinction between God's absolute and ordained power, especially as utilized in late medieval nominalistic theology, was an important background of Reformation theology. In his Disputation on Indulgences, Luther seems to espouse a theory similar to Ockham and Gabriel Biel, who followed Ockham in this matter.[37] According to Luther as well, the conservation of a thing is the same as its continuous creation (*concervatio rei est eius continuata creatio*). Creation, according to Luther, is 'to make constantly new' (*creare est semper novum facere*). The expression *semper* obviously means God's continuous creation.[38] It follows that all beings are at every moment immediately dependent on God, and every being in itself is purely *nihil*, nothingness. All being is continuously coming out (*semper in fieri*)[39] of nothingness and non-being, *ex nihilo*, to being by virtue of God's creative conservation.[40]

---

36 *Quaestiones in librum secundum Sententiarum. (Reportatio)*, Lib. II, q.3–4, OT V, 65;14–22. See also, Quodlibet VII, q.1, OT IX, 703–706. On whether Ockham should be considered a skeptic, see Marilyn McCord Adams, 'Was Ockham a Humean about Efficient Causality?' *Franciscan Studies*, 39 (1979), 5–48; Marilyn McCord Adams, *William Ockham. Vol. I–II.* (Notre Dame, Indiana 1987), 784–795; and Volker Leppin, 'Does Ockham's Concept of Divine Power Threaten Man's Certainty in his Knowledge of the World,' *Franciscan Studies* 55 (1998), 169–189.

37 In his commentary on the Sentences, Biel accepts Ockham' solution, see Gabriel Biel, *Collectorium circa quattuor libros Sententiarum. Liber secundus*, eds. W. Werbeck & U. Hofmann (Tübingen 1984), Lib.II, dist. 11, q.2; 25:15–22.

38 Martin Luther, *Resolutiones disputationum de indulgentiarum virtute (1518)*. Kritische Gesamtausgabe (Weimarer Ausgabe=WA), 1. Band. Weimar 1883–; WA 1, 563:6–13. Cf. David Löfgren, *Die Theologie der Schöpfung bei Luther* (Göttingen 1960) 37–45; Sammeli Juntunen, *Der Begriff des Nichts bei Luther in den Jahren von 1510 bis 1523* (Helsinki 1996), 167–174. The same idea can also be found in Luther's marginal notes on Lombard's Sentences (1510/11). WA 9, 66:29–35 and his lectures on Genesis (1535–1545). WA 43. 233:22–25.

39 The expression is the same that Aquinas uses while criticizing Islamic occasionalism in the *Summa contra gentiles*. See, 40, fn. 24.

40 In his early lectures, *Dictata super Psalterium* 1513–15, Luther explicitly represented the theological idea of the covenant or sine qua non causality, developed in the via moderna. WA 4. 261:32–39 See Alister McGrath, *The Intellectual Origins of the European Reformation* (Oxford 1987), 108–121. On the causality of sine qua non, see William J. Courtenay 'Covenant and Causality in Pierre d'Ailly,' *Speculum* XLVI (1971). In Courtenay, *Covenant and Causality in Medieval Thought*, IX, 110–119. After Luther rejected this idea of causality, and the Aristotelian-Thomistic theory as well, it is natural to ask in which direction his thinking was going? – According to *Dictata super Psalterium* the Scriptures use the term 'substance' only in a metaphorical and grammatical meaning, not in the physical or philosophical meaning. According to Luther, however, the Biblical use is more proper. The proper use of the term 'substance' is based only on the temporal continuity of the external qualities of a thing. WA 3. 419:25–27;35–38. The expressions simul and semper, repeatedly used by Luther in his Lectures on St. Paul's letter to Romans (1515–1516), are most plausibly interpreted as moving from the idea of covenant causality of the via moderna towards theological occasionalism, to a view of God's immediate

The idea of divine omnipotence and absolute power has been viewed as a central starting-point in Calvin's theology, indeed, even its trademark. Calvin's doctrine of the relationship between God and the natural world is more systematic, and has a more important role than that of Luther. For Calvin, the divine power is everywhere immediately present in nature, and nothing is based in an ultimate ontological sense on secondary causes. There is no autonomous activity and finality in nature that is not sustained by divine sovereign power. Physical reality is an instrument of God by which he acts as he wills and which he directs according to his own aims. For Calvin, no proper causal force in the world exists except the divine will that decrees secondary causes.[41]

Both Luther and Calvin denied Aristotelian-Thomistic essentialism and the idea of natural finality implied by it, understanding nature as a passive recipient of divine influence. There are no independent powers and faculties in nature or in man to co-operate with God's activity. Both considered the Aristotelian-Thomistic approach theologically pernicious. The two main streams of the Reformation, however, had different after-effects. Luther rejected the covenant model

---

causal influence. – 'Ideo simul sum peccator et iustus',...Divi Pauli apostoli ad Romanos Epistola. WA 56,70:9–10; 347:8–9; 441:21–442:17. From the point of view of Aristotle and Aquinas, Luther's semper and simul predications are contradictory; no entity can at same time be potential and actual, and always in the state of becoming. – On Luther's way of using the term nihil, see Juntunen, *Der Begriff des Nichts bei Luther.* – It is especially significant that Descartes says in the Discourse on the Method, that 'it is certain, and it is an opinion commonly accepted among theologians, that the act by which God now preserves it [the world] is just the same as that by which he created it.' René Descartes, 'Discourse on the Method of Rightly Conducting Reason and Seeking The Truth in the Sciences,' in: *The Philosophical Writings of Descartes. Vol. I* (Cambridge 1985), V 45, 133. – The atomistic conception of time is also closely related to occasionalism. Descartes considers that no natural thing has in itself a capacity to move from one instant of time to another instant of time. René Descartes, 'Meditations on First Philosophy,' in: *The Philosophical Writings of Descartes. Vol. II,* trans. J. Cottingham et al. (Cambridge 1984), III Meditation 49, 33–34. Some scholars, like N. Kemp Smith, do not regard Descartes as an occasionalist even though he represents a kind of voluntaristic creatianism and holds to the atomistic conception of time. Occasionalism is usually connected with the name of Nicholas Malebranche (1638–1715). See, Nicholas Jolley, *Light of the SoulM Theories of Ideas in Leibniz, Malebrance, and Descartes* (Oxford 1990), 99–113. Malebranche's occasionalism is a radicalized version of Descartes's argument in the 'Third Meditation,' cited above, where he presents his idea of the continuous creation (creatio continuata). Jolley, *Light of the Soul,* 99. See also Harry G. Frankfurt, 'Continuous Creation, Ontology of Inertia, and the Discontinuity of Time,' in: Harry G. Frankfurt, *Necessity, Volition and Love* (Cambridge 1999), 55–70; Margareth J. Osler, *Divine will and the mechanical philosophy: Cassendi and Descates on contingency and necessity in the created reality* (Cambridge 1994), 146–152.

41 Calvin does not fully deny an idea of universal providence. According to him, however, it is incorrect to use the idea to obscure the main importance of the special providence (specialis providentia) which concerns singular things. Special providence does not concern only a few single events but all singular things. Jean Calvin, *Institutiones Christianae religionis 1559. Johannes Calvini, Opera selecta. Vol. II* (Monachii 1928), Lib.I, cap. XVI, 194–195, 199.

of later medieval nominalism and its idea of causality of *sine qua non* which Biel, as a follower of Ockham, represented. In addition, Lutheran orthodox doctrine rehabilitated the Aristotelian view of nature and restored the basic difference between God's primary causality and natural secondary causes, a development which also took place to some extent in reformed theology. In contrast to Lutheran theology, Calvinism embraced the late medieval covenant model. Especially in the English branch of Calvinism, Puritanism, fascinating views of a new arising natural science along with the theological covenant model, gave religious motivation and legitimation to the utilization of the results of science. Science was seen as a secular but spiritually acceptable form of devotion.[42] The idea of divine absolute power, which minimized the autonomy of nature, makes understandable a mechanistic world view,[43] and partly explains why certain new scientific ideas were at first acknowledged by Lutheran theologians.[44] Copernicus's new theory of the mechanics of heaven, for example, aroused interest in Wittenberg. Copernicus himself gave the assignment of publishing his *De Revolutionibus* to a young mathematician Georg Joachim Rheticus[45] (1514–1574) from Wittenberg. The task was later given to an astronomer and theologian, Andreas Osiander, who wrote a cautious but in principle approving preface to Copernicus's book.[46] Rheticus's teacher was Erasmus Reinhold (1511–1574), who was one of the members

---

42   See Bernard I. Cohen, *Puritanism and the Rise of Modern Science: The Merton Thesis* (New Jersey 1990).
43   See Osler, *Divine Will and the Mechanical Philosophy*, 15–35; Gary B. Deason, 'Reformation Theology and the Mechanistic Conception of Nature,' in: David C. Lindberg and Ronald L. Numbers (eds.), *God and Nature: Historical Essays on the Encounter between Christianity and Science* (Berkeley 1986), 167–169.
44   Luther, Calvin and Melanchton have been mentioned as critics of Copernicus by Thomas Kuhn in his *The Copernican Revolution: Planetary Astronomy in the Development of Western Thought* (Cambridge, Mass. 1957), 196. It is true that Luther in his Tischrede criticized Copernicus's new theory. Hooykaas has shown that Calvin does not mention Copernicus at all. On the contrary, Calvin does not interpret the Genesis literally. According to Calvin's principle of accommodation, many divine truths are accommodated in human understanding. Scripture is not a book supernaturally revealed in natural things. Reijer Hooykaas, *Religion and the Rise of Modern Science* (Edinburg 1972) 121. See also John Dillenberger, *Protestant Thought and Natural Science: A Historical Interpretation of the Issue Behind 500-year-old Debate* (New York 1960), 28–49; Heiko Oberman, *The Dawn of the Reformation*, 183–187.
45   On Rheticus, see Reijer Hooykaas, *G.J. Rheticus' Treatise on Holy Scripture and the Motion of the Earth: With Translation, Annotations, Commentary and Additional Chapters on Ramus-Rheticus and the Development of the Problem before 1650* (Amsterdam 1984).
46   The role of Osiander has been in dispute. I follow here Oberman's convincing arguments. According to Oberman, it is true that Osiander in his unauthorized preface tried to belittle Copernicus's ideas by calling them 'hypotheses.' But to speak about the 'fraud' of Osiander is not correct. Copernicus himself used the term 'hypothesis' for his new ideas. Copernicus's letter to Pope Paul III contains significant parallels with Osiander's preface. Oberman, *The Dawn of the Reformation*, 189–192. Cf. Pierre Duhem, *To Save the Phenomena: An Essay on the Idea of Physical Theory from Plato to Galileo* (Chicago 1969).

of Melanchton's circle. Melanchton's son-in-law, Caspar Peucer, was likewise a student of Reinhold and Rheticus. Mathematical astronomy had until the end of the 16th century an important place in Wittenberg, influencing other universities in Germany, and also in Scandinavia.[47]

It would be too much to say that the conceptual changes of this period alone were sufficient reason to explain the birth of a technological interest in nature. Many external, social, political, and economic reasons have not been mentioned here. But one can state that, as internal factors, these changes were a necessary condition to prompt some of those central efforts which in the next centuries formed a conceptual basis for a technological form of life. This change in attitude towards nature presupposed a detachment from the Aristotelian-Thomistic categories of nature and substance as a teleological conception of nature. This notion was replaced by a concept of nature as passive and as outer directed. Without these conceptual chances it is difficult to imagine how the idea of harnessing nature to serve human needs could have come into the human mind.

One essential difference between the new natural science and medieval science was in the significance of mathematics in understanding the regularities of nature. This emphasis of mathematics also presupposed a disengagement from certain Aristotelian ways of thinking, though it also presupposed paradoxically a return to certain Platonic ideas. While repudiating Plato's conception of separate ideas, Aristotle had also cast off Plato's idea that mathematics represents, at best, intelligible forms which are reflected only incompletely in the changeable, physical world. From the Aristotelian point of view, quantities and other mathematically measurable forms are abstractions from accidental physical features of a thing and do not display the real nature and essence of a thing. Science concerns the substances of beings, and they are not reducible to essentially mathematical forms. The real objects of Aristotelian science are not quantities but qualities, especially, the substantial qualities of beings. A return to the Platonist view of the separateness of the ideas (i.e. numbers) from physical entities rendered possible the reduction of natural forms to numbers and geometrical forms. It also meant a return to a kind of aesthetic view of nature in which mathematics was seen as representing a fundamental universal harmony.[48] At the same time a different

---

47 See Westman 'The Copernicans and the Churches,' in: *God and Nature*, 76–113; Heiko Oberman, 'Reformation and Revolution,' in: *The Cultural Context of Medieval Learning*, 397–435.
48 'For Kepler, of course, the real qualities are those caught up in this mathematical harmony underlying the world of the sense, and which, therefore, have a causal relation to the latter. The real world is a world of quantitative characteristics only; its differences are differences of number alone. In his mathematical remains there is a brief criticism of Aristotle's treatment of the sciences, in which he declares that the fundamental difference between the Greek philosopher and himself was that the former traced things ultimately to qualitative, and hence irreducible distinctions, and was, therefore, led to give mathematics an intermediate place in dignity and reality between sensible things and the supreme theological or metaphysical ideas; whereas

kind of conception of matter from that of Aristotle and Aquinas was adopted. Matter was viewed as merely passive in itself, factually controlled and directed by external laws and forces, and no longer as an organic wholeness directed by the substantial forms potentially included in it.[49] The result was a mechanistic worldview in which the laws of nature were understood as external, separated from passive matter, but nevertheless as the laws which determine the course of nature.[50] Finally, this reduction of the laws of nature to mathematical laws, on which even God is dependent, given its obvious explanatory power, was apt to secularize scientific rationality so that natural laws became an independent model of explanation without necessary reference to God.[51]

---

he had found means for discovering quantitative proportions between all things, and therefore gave mathematics the pre-eminence.' Burtt, *The Metaphysical Foundations of Modern Physical Science*, 67.

49 'Rather than being a correlative term, always associated with form (as it was on either the Aristotelian or the Platonic concept), matter came to be regarded as having a "free-standing" status. This change brought with a new set philosophical problems and the reinterpretation of such traditional terms as "matter," "substance," "form," "quality," "cause," and "activity" in mechanical terms.' Osler, *Divine will*, 172.

50 One interesting topic in this discussion was the question whether mechanics is a theoretical (i.e. mathematical) or is it practical science, that is to say, does it concern thinking and ways of thought, or constructing things? Besides Aristotle's mechanics as such, the Pseudo-Aristotelian work called Problemata Mechanica, unknown to medieval authors, was much commented upon. The reason for this question was that even if the object of mechanics is matter, its methods, nevertheless, are by nature mathematical. Among the others, Alessandro Piccolomini (1508–1579) and Giuseppe Moletti (1531–1588), the latter a mathematician from Padova, took the position that mechanics is a contemplative mathematical science, subjected to geometry (according to the Aristotelian subalternation model of sciences). According to Moletti, the task of mechanics is to teach how the difficulties caused by nature can best be overcome by means of art (ars), using minimal force. He also rejected the disreputable etymology of the word mechanica, which was supposed to come from the Latin word moechus, adultery. Nor was it uncommon to compare the laws of nature with the positive laws of society, with the distinction however, that the former are mathematical whereas the latter are not. Heikki Mikkeli, *An Aristotelian Response to Renaissance Humanism: Jacopo Zabarella on the Nature of Arts and Sciences* (Helsinki 1992), 119–125.

51 Robert Boyle's (d.1691) attitude is in this respect a typical example. According to him, 'Nature plays a mechanician.' Mathematical and mechanical principles are the alphabets by which God wrote the world. As far as God himself 'played a mechanician' while creating the world, he also was dependent on those laws. Mathematical principles, like logical ones, as the ultimate truths, are above God and independent from revelation. Their explanatory power does not presuppose referring to God. Burtt, *The Metaphysical Foundations of Modern Physical Science*, 173. Boyle regards as central idea of Christianity the fundamental difference between God and creatures. The Bible, according to him, does not include any reference to secondary or cooperative causes, but describes everything as effects of God's immediate causality. Deason sees in Boyle's view an influence of the Protestant religion, where the emphasis of God's sovereign and absolute power is central. Deason 1986, in *God and nature*, p. 180. On Boyle, see also Reijer Hooykaas, *Boyle: A Study in Science and Christian Belief* (Lanham, Md. 1997).

4. On the Relationship between Metaphysics and Physics

Olli Koistinen (Turku)

1. Introduction

Some physicists, especially those advocating what is called the Theory of Everything, seem to hold that certain perennial problems in metaphysics can be solved in a fully developed physical theory. I have a modest aim in this paper. What I want to do is to give some sort of characterization of metaphysics and the problems it deals with, and I hope that this alone is enough to show that metaphysical problems cannot be solved in physics.

In the first section of this paper, Aristotle's conception of metaphysics is shortly displayed, and an ambiguity in his conception of metaphysics is paid attention to. In the second section, Immanuel Kant's idea of metaphysics as the system of synthetic a priori knowledge is taken seriously. It seems that great dogmatic metaphysicians Descartes, Spinoza and Leibniz held that the ultimate nature of reality, and that important truths about it, can be known without empirical justification. Now, it seems that this kind of conception of metaphysics may be the reason why some feel a tension between the claims of physics and the claims of metaphysics. However, a look at the principal claims of the most important work of that era, viz. Descartes's *Meditationes de prima philosophia,* shows that Descartes was more concerned to make physics possible than to argue for any particular theory in physics. In the third section, Kant's view that synthetic a priori knowledge about the world is impossible, is considered. Kant, then, rejected metaphysics in the sense the rationalists were pursuing it. But Kant's attitude towards physics can also be seen somewhat destructive. Even physics is unable to tell important truths about the reality as it is independently of any observer or thinker. In the fourth section, certain problems in general metaphysics or ontology are considered. It seems that these problems, typical of the current metaphysical inquiry, have nothing to do with physics.

2. Aristotle's First Philosophy

Aristotle called *first philosophy* the science that was later dubbed as metaphysics. The object of first philosophy is being qua being:

There is a branch of knowledge that studies being qua being, and the attributes that belong to it in virtue of its own nature. Now this is not the same as any of the so-called speculative sciences, since none of those enquires universally about being qua being. They cut off some part of it and study the attributes of this part—that is what the mathematical sciences do, for instance. But since we are seeking the first principles, the highest causes, it is of being qua being that we must grasp the first causes.[1]

Thus, metaphysics seems to be the most abstract and general of all sciences. A science which investigates something that is somehow associated with everything there is – viz. being.

However, later in the *Metaphysics* the first philosophy is characterized in less general terms. Aristotle tells that first philosophy should be seen as the science in which the nature and existence of an unchanging separate substance, i.e. God, is considered. It certainly comes as no surprise that commentators have found it problematic how to reconcile these different characterizations of what should be one and the same science. J.L. Ackrill[2] points that an attempt to unify these distinct conceptions occurs in E.1.1026a10:

There are therefore three kinds of theoretical philosophy: mathematical, natural and theological. (I call the study of the changeless and eternal 'theology' because this is obviously the category into which divine being falls.) Now the highest science must deal with the highest kind of thing. So, while the theoretical sciences are to be preferred over others, this theology is to be preferred over the other theoretical sciences.

One may raise the question whether first philosophy is universal or deals with just one kind of thing. If there were no substance other than those formed by nature, natural science would be the first science; but if there is a changeless substance, knowledge of this must be prior and must be *first* philosophy – and universal just because it is first. And it will belong to it to consider being qua being, both what it is and the attributes that belong to it qua being.

However, I agree with Ackrill[3] that Aristotle does not give a satisfactory account of how general metaphysics could be reduced to theology. Even if it were true that God is the explanation and cause of all changes and changing things, 'it does not follow that knowledge of God will *include* knowledge of objects and

---

1 Aristotle, *Metaphysics*, in: J.A. Smith and W.D. Ross (eds.), *The Works of Aristotle translated into English* Vol.VIII (Oxford 1910–52), G.1. 1003a21.
2 J. L. Ackrill, *Aristotle: The Philosopher* (Oxford 1981), 118.
3 Ackrill, *Aristotle*, 119.

changes, or that theology will itself be concerned to study the attributes of being qua being.'

Thus, it seems that metaphysics was born ambiguous. It may mean either the general investigation into the nature of being itself, or it may mean the investigation into the nature of God. For the rationalists whose metaphysics was closely connected with the new physics of the early modern era, metaphysical knowledge, to use the terms later invented by Immanuel Kant, was closely connected to a priori synthetic knowledge about reality.[4]

3. *Descartes's Metaphysics*

Maybe the most important work written in modern philosophy is René Descartes's *Meditationes de prima philosophia*. Descartes characterizes its six meditations as follows:

Meditatio I: De iis quae in dubium revocari possunt (What can be called to doubt)

Meditatio II: De natura mentis humanae: Quod ipsa sit notior quam corpus (The nature of the human mind, and how it is better known than the body)

Meditatio III: De deo, quod existat (The existence of God)

Meditatio IV: De vero et falso (Truth and falsity)

Meditatio V: De essentia rerum materialium et iterum de Deo, quod existat (The essence of material things, and the existence of God considered a second time)

Meditatio VI: De rerum materialium existentia et reali mentis a corpore distinctione (The existence of material things, and the real distinction between mind and body)

It seems that for Descartes metaphysics, or first philosophy, has not much in common with Aristotle's view of metaphysics. What Descartes's metaphysics shares with Aristotle's metaphysics is the concern with God's nature and existence. However, general metaphysics seems to be lacking completely from *Meditationes*. So why does Descartes, then, emphasize in the title of the book that it is a treatise about first philosophy? Isn't it more like a treatise in foundationalistic epistemology than in metaphysics? Now, I believe that one possible explanation of this is

---

4   About the relation between these two, at least apparently distinct, conceptions of metaphysics in Aristotle, see Joseph Owens, *The Doctrine of Being in the Aristotelian Metaphysics* (Toronto 1951).

that Descartes thought that, in fact, his work does contain the foundations of his version of the new physics, and thus literally is metaphysics. At least, Descartes wrote to Mersenne (28 January 1641) as follows:

> I may tell you, between ourselves, that these six meditations contains all the foundations of my physics. But please do not tell people, for that might make it harder for the supporters of Aristotle to approve them. I hope that readers will gradually get used to my principles and recognize their truth, before they notice that they destroy the principles of Aristotle.[5]

What is distinctive in Aristotelian concept formation is that all concepts are derived from an experiential touch with the world. As Descartes's scholastic predecessors put it: 'nihil est in intellectu quod non prius fuerit in sensu'. However, the problem with Aristotelian concept formation, from the perspective of the new physics, is that the new physics seems to exlude all so-called sensible qualities, such as colours and odours, from its laws and explanations. Thus, provided that physics gives the best account of the nature of mind-independent reality, it seems that the true story of the world is not the one told by our senses but a story that uses intellectual concepts instead.The first and second meditations can be seen to pave way for this. Let us see how.

In the first meditation, 'De iis quae in dubium revocari possunt', Descartes problematizes the trustworthiness of the information based on sensual experience as follows:

> Whatever I have up till now accepted as most true I have acquired either from the senses or through the senses. But from time to time I have found that the senses deceive, and it is prudent never to trust completely those who have deceived us even once.[6]

For Descartes, sense perception is, in a sense, inherently misleading, and the justification of this is one of the main aims of the first meditation. Now, I believe that this kind of assumption of senses as not being a reliable fundamental criterion of reality is what is shared at least by those physicists who endorse scientific realism, and that Descartes's philosophical argumentation can be seen to be intended to help the physicists.

---

5   René Descartes, *Letter to Mersenne* (28 January 1641), in: John Cottingham, Robert Stoothoff, Dugald Murdoch and Anthony Kenny (eds.), *The Philosophical Writings of Descartes* Vol.III (Cambridge 1984), 173 AT III 298.
6   Descartes, *Meditations on First Philosophy,* in: John Cottingham, Robert Stoothoff and Dugald Murdoch (eds.), *The Philosophical Writings of Descartes* Vol.II (= CSM II) (Cambridge 1984), 12 AT VII 18.

In the second meditation, Descartes presents his argument that he cannot doubt his own existence even when he doubts the existence of everything that is material. This leads Descartes to conclude that his knowledge of mind is better than knowledge of anything that is corporeal. In the second meditation, Descartes already concludes that he is a substance whose essence consists in thought, so starting his argument for mind-body dualism which he concludes in the fifth meditation. Descartes's dualism has been ridiculed and despised. However, it seems to me that Descartes's argument for dualism, in spite of its initial implausibility, is a remarkable achievement and given his theoretical view of identification and individuation it is not a non-sequitur but a valid argument.[7] Now, for the theorists of everything, dualism, of course, poses a threat. But I do not understand how any argument using premises that are derived solely from physics could, without making any philosophical commitments, show that dualism is false. I will return to the issue of dualism later. In the second meditation, Descartes also gives his famous proof that the extension of matter does not consist of sensible qualities and that matter even does not possess such qualities. Descartes presents the proof as follows:

> Let us consider the things which people think they understand most distinctly of all; that is the bodies which we think we touch and see. I do not mean bodies in general – but one particular body. Let us take, for example, this piece of wax. It has just been taken from the honey comb; it has not yet quite lost the taste of the honey; it retains some of the scent of the flowers from which it was gathered; its colour, shape and size are plain to see; it is hard, cold and can be handled without difficulty; if you rap it with your knuckle it makes a sound. In short, it has everything which appears necessary to enable a body to be known as distinctly as possible. But even as I speak, I put the wax by the fire, and look: the residual taste is eliminated, the smell goes away, the colour changes, the shape is lost, the size increases; it becomes liquid and hot; you can hardly touch it, and if you strike it, it no longer makes a sound. But does the same wax remain? It must be admitted that it does; no one denies it, no one thinks otherwise. So what was in the wax that I understood with such distinctness? Evidently none of the features which I arrived at by means of the senses; for whatever came under the taste, smell, sight, touch or hearing has now altered – yet the wax remains.[8]

---

7 For a fuller defense of this view, see Olli Koistinen and Timo Kajamies, 'Descartes on Non-Identity and Dualism,' in: Tuomo Aho & Mikko Yrjönsuuri (eds.) *Norms and Modes of Thinking in Descartes* (Helsinki 1999).
8 CSM II, 20 AT 30.

This example quite convincingly shows that the wax cannot be identified with the sensible qualities associated with it at a given moment of time. If it could, it couldn't survive the loss of those qualities associated with it at some moment of time. It is evident that Descartes's experiment can be extended to cover all material bodies, which gives good evidence for the view that the essence of matter does not lie in sensible qualities. What, then, is the essence of matter? Descartes suggests that it has something to do with shape. However, it cannot be any particular shape a matter might take on, because, as the wax example shows, also the shape of a body can change. Descartes concludes from this that the essence of the wax and, by a parity of reasoning, of all bodies is their *capability* of being extended in many more different ways than a finite being can imagine. Now, this capability is nothing that can be perceived by senses or can be found out by sense based imagination but is something that can be perceived by the mind alone:

> I would not be making a correct judgment about the nature of wax unless I believed it capable of being extended in many more different ways than I will ever encompass in my imagination. I must therefore admit that the nature of this piece of wax is in no way revealed by by my imagination, but is perceived by the mind alone. ...[A]nd here is the point, the perception I have of [the wax] is a case not of vision or touch or imagination – nor has it ever been, despite previous appearances – but of purely mental scrutiny; and this can be imperfect and confused, as it was before, or clear and distinct as it is now depending on how carefully I concentrate on what the wax consists in.[9]

When the objects of physics have been transformed into objects of intellectual cognition, the metaphysical work for the new physics has in a certain sense been accomplished. Physics cannot be criticized on the basis that the picture offered by it differs from the ordinary sense-painted picture of the world. In fact, what has been shown by Descartes is that premises that are quite commonsensical lead to the view that the world is not what our senses tell it to be.

In the third as well as in the fifth meditation, Descartes purports to prove the existence of God and is, hence, engaged in metaphysical investigation as conceived by Aristotle. It is quite interesting why Descartes presents two *different* arguments for the existence of God. According to the first argument, God's existence follows from the fact that we have an idea of an infinite being and from the further premise that there must be at least as much reality in the cause as there is in the effect. Thus, the idea of God must be caused by an infinite being, i.e. by God himself. According to the second argument, existence is no more separable from the essence of God than the concept of a valley is from the concept

---

9   CSM II, 21 AT 31.

of a mountain. This means that a sentence which denies the existence of God is contradictory in the same sense as the sentence which affirms the existence of a square triangle. Now, it may be that the reason why Descartes presented an alternative proof in the fifth meditation is that the causal principle required by the first proof is not obvious a priori and resembles empirical laws of physics. The latter argument clearly is thoroughly non-empirical. Thus, in proving the existence of God, who for Descartes guarantees the immutability of physical laws, Descartes can be seen to be cautious in not confusing physics and metaphysics. Maybe physics can be seen to be relevant to the so-called a posteriori proofs of God's existence but because we will shortly see that for Descartes reliance on a posteriori premises is dependent on the existence of God, he wouldn't appeal to such premises. Thus, it seems that Descartes would say that physics is irrelevant to the claims concerning God's existence.

In the fourth meditation, Descartes lays down a criterion of truth according to which the clarity and distinctness of an idea is a sure sign of its truth. I believe that this criterion is something researchers of theoretical physics should greet with joy. For Descartes, sense perception is not clear and distinct but these attributes are applicable only to intellectual perception. But the objects physics deals with are, as we have seen, objects of intellectual perception, and therefore possible objects of clear and distinct perception, and, thus, possible objects of knowledge. The fourth meditation should be seen as doing a great favour to the physicists in making a contribution to the possibility of physical knowledge. Moreover, I am unable to see how the justification of a truth criterion could be done from premises that contain nothing but concepts of physics. A criterion of truth for the basic principles of physics can quite plausibly be called *meta*physical.

The sixth meditation nicely completes Descartes's agendum of making physics possible. In this meditation, Descartes purports to argue that there really is a subject matter for physics; i.e. that the concepts of physics have an extramental reality and are not figments of imagination. Descartes does this by appealing to the fact we have a quite strong propensity to believe that our sensory beliefs are caused by objects outside us. Because Descartes believes to have shown that God, who has created us, exists, he has to believe that this very propensity to believe in the reality of the external world comes from God. Thus, if there were no external world, God would be a deceiver in letting us to believe in the existence of such extramental reality. Now, it seems to me that the proof of the existence of the physical world cannot be given inside physics – also that is a metaphysical task.

As a sort of conclusion to this section the following might serve: In Descartes's metaphysical treatise there are no questions or problems that could be solved in physics. However, that treatise should not be seen as irrelevant to the claims of physics. Descartes shows that physics with its domain of intellectual objects can be shown to be commonsensical. He also shows that physics can reach the truth about natural objects by appealing to the existence of God which he proves in two ways. By dividing the world into matter and mind Descartes can also be seen to shown that physics does not undermine the requirements of morality. Even though there is much in Descartes's argumentation that is not quite convincing, he quite admirably shows what consists in the metaphysical justification of physics.

4. *Kant and Physics*

Spinoza's and Leibniz's philosophies can be seen as further developments of Descartes's philosophy, and also Immanuel Kant began to philosophize in the way Descartes and, more importantly, Leibniz did. However, Kant got a bad hangover from his early metaphysical and epistemological commitments. Rationalists, in philosophizing in mathematical manner, were in problems because they were relying on reason alone and Kant did not understand how reason can tell something informative (i.e. how is synthetic a priori knowledge possible) about something that is independent of reason. It even seemed to Kant that the rationalists did not succeed in proving the existence of the external world. And the lack of such proof was according to Kant the scandal of philosophy. Kant also thought a priori arguments for the existence of God were impossible and we have already seen the importance of God in Descartes's metaphysics. To recover from that most unpleasant condition Kant begins to develop a new way of explaining experience. And in *Kritik der Reinen Vernunft* Kant presents his doctrine of transcendental idealism. In transcendental idealism which is consistent with Kant's version of empirical realism it can be shown that physics is consistent with morality and religion. However, the semantico-conceptual underpinnings of transcendental idealism entail that even though physics as a science is possible, metaphysics in its *traditional sense* is impossible. For Kant, metaphysics became the doctrine about what can be known of reason and experience a priori. In metaphysics, a priori knowledge was still the aim, but the object of that knowledge was reason and experience itself.

The basic tenet of Kant's transcendental idealism is the ideality of space and time. Space and time are forms of sensibility (Sinnlichkeit) which means that all experience shows its objects in space and in time. In being conditions of experience, the concepts of space and time, instead of being empirical, are a priori. These forms of sensibility connect objects together so that objective experience becomes possible. These objects, which the forms of sensibility tie together, are for Kant strictly speaking appearances of 'things as they are in themselves'. Kant believes that because space and time are ideal pertinent metaphysical problems disappear. It makes no sense to ask whether space and time are infinite or finite, because strictly speaking there is no space neither time. Thus, all kinds of questions about the totality of mind-independent things are senseless and classical metaphysics with its concern with the origin of everything is asking these illegitimate questions about totality.

As we just said, for Kant, the objects of experience are appearances of 'things as they are in themselves'. Kant's reasons for postulating such a realm of 'things as they are in themselves' are twofold: (i) we have a strong tendency to believe that in some of our experiences we are being acted on. This kind of passive experience happens in perception. But if we are being acted on, there must be something that acts on us. And the things acting on us are things as they are in themselves. (ii) The objects of experience are subject to strict causal laws. But morality requires contracausal freedom. Thus, even though all objects of experience are causally determined, as things in themselves, human beings can be free and morally responsible.

For Kant, physics deals with objects which are conceived as existing in space, and for him, the task of metaphysics is to show that physics does not study mind-independent reality, but that it confines to objects that exist only in experience. Now this can be seen to have some implications to physics itself. Because Kant believes that any object must be experienced as in space and time and that objects as experienced must conform to strict causal laws, it follows that physics must necessarily be deterministic. Even physics does not investigate the nature of things as they are in themselves. On the other hand, metaphysics cannot show anything but that all objects must have sufficient causes, it does not tell anything about the particular causal laws. Physical investigation, Kant says, does not need metaphysics to reach its principles but a proper understanding of the law-governedness of the experienced world requires metaphysical knowledge:

> [M]etaphysics does not contribute to the extension of empirical principles, of the science of empirical physics: its cognition is wholly unnecessary with respect to physics, where the principles of metaphysics are put wholly to the side, and one starts from settled appearances, and the principles derived from that are adequate to explain everything from them. ...It is also certain that every man is occupied with his own metaphysics, and of that there is truly no doubt, metaphysics must be explained; but everything depends on the method with which it is treated. Therefore all despisers of the metaphysical sciences punish themselves in that they despise science itself and yet treasure their own method for thinking about supernatural truths. In short, no human being can be without metaphysics.[10]

Thus, even Kant seemed to keep metaphysics and empirical physics separate.

---

10 Immanuel Kant, *Metaphysik Vigilantius,* in: Karl Ameriks and Steve Naragon (eds.), *Lectures on Metaphysics* (Cambridge 1997), 420.

5. General Metaphysics and Ontology

Aristotle's idea of the first philosophy as a science that studies being qua being marked the start of that part of philosophy that was labeled ontology by Christian Wolff. In contemporary philosophy, ontology and metaphysics have almost become one and the same. In ontology such questions as the nature of particularity and generality are discussed, and in this study the general commitments of human experience and inference are investigated. Even though it is not so easy to give an exhaustive and illuminative definition of ontology, it is rather easy to show, with the help of examples, the nature of ontological enquiry. I will use as examples the following problems: (i) the problem of the universals; (ii) the problem of the nature of a particular thing; (iii) the problem of the identity of properties.

The problem of the universals is this. Suppose that both

1. Mary is wise

and

2. Peter is wise

are true. Now it seems that from the conjunction of 1 and 2 it can be concluded that

3. There is some common element (viz. wisdom) that belongs both to Mary and Peter.

The problem that 3 generates is: how can one thing (i.e. wisdom) be in two distinct things when they do not share it as they might share a piece of bread?

Now, the so called Platonists would say that in fact the analogy of sharing a piece of bread is not at all so bad. Wisdom is a universal property or concept and such things do not have spatial existence. In a sense both Mary and Peter can 'share' this property. They both can be related to it. But what is the nature of this relation and what is the nature and way of existence of nonspatiotemporal concepts? These are horrible problems. Can they be solved in physics?

The problem of the nature of a particular thing is this. Consider a blue ball. It seems that for it to exist there must be something that has the properties of being spherical and being a ball. But what is the thing that is supposed to have the properties of being a ball and being spherical? More generally, what is the particular thing contrasted to its properties? What remains when all the properties of a thing are stripped away from it? Can this problem be solved in physics?

The problem of the identity of properties is this. When an ontological kind is introduced, an identity criterion for that kind must be given; i.e. some sort of principle must be given which says when two expressions, purporting to refer to entities assumed to belong to that kind, specify the same entity and when different. For material things such identity criterion could be the similarity of spatiotemporal location but what could be a reasonable criterion for properties? A plausible initial suggestion is the following:

(C1) Properties are identical iff the sets of their instances (i.e. their extensions) are identical.

However, this suggestion is problematic because there can be properties that are, as it were, accidentally co-instantiated. For example, W.v.O. Quine has suggested that the properties of having kidneys and having a heart have identical extensions, i.e. they have the same instances. But it is clear that having a heart and having kidneys are not identical properties.

It has been suggested that (C1) should be strengthened by the necessity operator. Thus, the modified criterion would be:

(C2) Properties are identical iff they necessarily have the same instances.

According to (C2) having kidneys and having a heart would not be identical properties because it is conceivable that there are creatures who do have a heart but who lack kidneys. However, (C2) generates new problems. What does 'necessarily' mean? In the semantical developments of modal logic attempts have been made to analyze necessity with the help of the concept of possible world. A proposition is said to be necessary when it is true in all possible worlds. However, the problem with this suggestion is that maybe it is not at all easier to understand the expression 'true in all possible worlds' than 'necessarily'. But, of course, the *analysans* should be better known than the *analysandum.* Moreover, according (C2) all trivial properties, i.e. those that are necessarily exemplified by everything there is such as being self-identical and being something or other, would be identical. Roderick Chisholm[11] has suggested that (C2) should be replaced by the following:

(C3) Properties are identical iff they necessarily have the same instances and they are necessarily conceived through each other.

---

11 Roderick Chisholm, *Person and Object* (London 1976).

According to (C3), there can be distinct trivial properties but (C3) does have problems of its own. What does it mean that something is necessarily *conceived* through something else? Doesn't this somehow make the identity of properties mind relative? Be this as it may, it, however, seems clear to me that physics cannot provide us with any identity criterion of properties at all. But such an identity criterion is needed when, for example, questions about the identity between mental states and brain states are considered. A materialist, or anyone believing in some kind of fundamental basic stuff, cannot be taken seriously if her or his argumentation does not rely on a plausible identity criterion of states or properties.

Sometimes physicians accuse philosophers of endorsing a somewhat naive common sensical substance-property ontology. This kind of picture of the world has been characterized by C.S. Peirce as the house of the Stagyrite.[12] Is it, then, possible that this kind of substance property is based on narrow sensual touch with the world? Could theoretical knowledge about the fundamental sructure of reality also show the falsity of the substance-property ontology? I believe that it could, if basic ontological concepts are abstracted from experience. However, I have serious doubts about the empiricality of ontological concepts. My Kantian, or Spinozistic, point is as follows. Thinking, understanding and experiencing have necessarily a judgmental character. What it means is that in these mental acts something is by necessity *predicated* of something else. Now, the concepts of substance and property are conceptually connected to this affirmative nature of thought. Properties are those entities that are predicated of substances. If this is true, then an ontology which could replace substance and property by some other ontological kinds would mean saying farewell to all experience and thinking. Where there is thought there is substance and property; thus rejection of those categories is really inconceivable. But even though I believe that substance-property ontology is indispensable this does not entail any view about the nature and number of these entities. For example, Spinoza's ontology, where there is only one substance and where all ordinary individual things are nothing but properties of this only substance, is as good a specimen of substance-property ontology as Aristotle's ontology where all individual things are distinct substances.[13]

---

12 I owe this to Kenneth Olson's *An Essay on Facts* (Stanford, CA 1987), 19. Olson gives the following reference to Peirce: C. Hartshorne and P. Weiss (eds.), *Collected Papers* Vol.1 (Cambridge, Mass. 1931–58), vii.

13 I want to thank Arto Repo for helpful discussions and Professor Eeva Martikainen for inviting me to this conference.

5. The Concept of Infinity in Kant and Hegel

Jussi Kotkavirta (Jyväskylä)

1. Introduction

In his critical philosophy Kant's general strategy is to redefine the foundations as well as the limits of metaphysics by a novel synthesis of rationalistic and empiricistic lines of thought. This project of synthesis is also true of his ideas concerning the notion of infinity, which has several important roles in Kant's architectonic construction and which gives a good perspective on studying that construction. The notion of infinity and the problems connected with it are especially significant in *Kritik der reinen Vernunft*.[1] In this work Kant opens his critical path between metaphysical scepticism and dogmatism, also redefining the traditional views on infinity. In his *Kritik der praktischen Vernunft*[2] Kant develops and justifies his practical metaphysics with potent ideas about ourselves as infinite moral persons. Finally, in his *Kritik der Urteilskraft*[3], Kant's discussion of the sublime contains in particular important reflections about infinity and its relation to the human mind. I will confine my discussion to the three critiques and leave the other texts mostly unconsidered, although they certainly contain, as does Kant's late volume on religion, important thoughts on infinity.

Kant is convinced that reason and freedom are truly infinite, though they can never be objects of human experience. True infinity is not related to the world of space and time, which is neither finite nor infinite. Our knowledge of nature is confined to the limits of possible experience, which for Kant implies that the traditional notions of the metaphysical infinity of the world, of the human soul, and of God, have to be subjected to critical analysis and in the main rejected. This is demonstrated in the mathematical antinomies of KrV, of which I will briefly discuss the first two. Even though we cannot have empirical knowledge about the infinitude or finitude of the world as such, transcendental reflection may reveal to us that our experience operates with notions of infinity. Both time and space as forms of sensible intuition and our judgements of understanding make use of an in-

---

1   The quotations are taken from the translation of Werner S. Pluhar (Indianapolis 1996)(= KrV). I refer to the page numbers of the two editions A and B of 1781 and 1787 according to the Akademieausgabe (= AA).
2   The quotations are taken from the translation of Mary Gregor in: Immanuel Kant, *Practical Philosophy* (Cambridge 1996) (= KpV). I refer to the page numbers of the Akademieausgabe.
3   The quotations are taken from the translation of Werner S. Pluhar (Indianapolis 1987) (= KU). I refer to the page numbers of the Akademieausgabe.

finity. This notion is a mathematical infinity, however, which points only towards the truly metaphysical dimensions of infinity, i.e. towards reason and freedom. Kant's transcendental idealism is built on the view that our experience of nature can never reach nature in itself but is limited to its own appearances or presentations. It is well known that Hegel is much more deeply rationalist and realist in maintaining that there is no such gap between reality in itself and its appearances to us. Hegel's general strategy is to argue that reality in itself and its rational appearance to us are ultimately the same thing and identical with each other. This also entails that metaphysical infinity is not beyond the reach of our knowledge. Quite the contrary, it is constitutive of our rational knowledge of reality. Hegel follows most notably Spinoza in his formulation that the object of such rational knowledge is God, i.e., true infinity. Consequently, Hegel rejects the Kantian limits and oppositions regulating our insight into the metaphysical infinity of the world and also of ourselves.

Hegel had formulated his ideas concerning the notion of infinity already in Jena during the very first years of 19th century. His extended essay *Glauben und Wissen*[4] from the year 1802/03 is particularly important in the context of my discussion, because there Hegel discusses thoroughly and critically Kant's transcendental idealism, defending his own dialectical monism with its very different views on infinity and our capacities for knowing it. Actually, in most of Hegel's writings, as well as in the writings of other post-Kantian idealists in Germany, infinity and its various dimensions or aspects are a major theme. Most often Hegel uses the term 'absolute' for the infinite, or present the infinite as one aspect of the absolute. His later logic contains very succinct formulations of the notion and its various aspects. I will use in particular its later and shorter formulation found in the *Enzyklopädie der philosophischen Wissenschaften* (1830)[5]. My discussion will, for the most part, be interpretative, and I will limit my constructive remarks to the Kantian and Hegelian frameworks. Thus I will not try to judge their theories of the infinite in the light of more recent ideas on the topic.

2.

The world out there, 'the starry heavens above us', which Kant famously refers to in the conclusion of the *Kritik der praktischen Vernunft*, is according to him experienced as a metaphysically infinite and absolute whole. It '(...) begins from the place I occupy in the external world of the sense and extends the connection

---

4   The quotations are taken from the translation of Walter Cerf and H.S. Harris (Albany 1977) (=GW). I refer also to the page numbers of Hegel's *Gesammelte Werke* (Hamburg 1968), Bd.4.
5   The quotations are taken from the translation of T.F. Geraets, W.A. Suchting, H.S. Harris. G.W.F. Hegel, *The Encyclopaedia Logic* (Indianapolis 1991).

in which I stand into the unbounded magnitude with worlds upon worlds and systems upon systems, and moreover into the unbounded times of their periodic motion, their beginning and their duration'.[6] The starry heavens, Kant continues, 'annihilates, as it were, my importance as an *animal creature*', being thus a vivid demonstration of 'our finitude in the cosmic world as a whole.' It is important to realise what Kant is actually doing here, namely that he is employing an intelligible, 'rational', or metaphysical concept of world, which he distinguishes from the 'empirical' concept of world.[7] Thus the starry heavens remains, beside 'the moral law within me', as the two paradigms of true infinitude.

Kant suggests that an experience of the infinite heavens is sublime. As such, it exceeds the limits of our possible experience, and only human reason can control the mind's confrontation with such an object. According to Kant's transcendental idealism, such an experience cannot constitute knowledge of the world or of its infinitude. As humans we are finite beings for whom this world appears only through our receptivity, i.e., through our sensible intuitions.[8] Our knowledge of nature is confined to the limits of our possible experience. In our receptivity we are, according to Kant, metaphysically finite beings incapable of knowing the world as it in itself. Consequently, 'intellectual intuition seems to belong solely to the original being, and never to a being that is dependent as regards both its existence and its intuition an intuition that determines that being's existence by reference to given objects'.[9] Our knowledge of the external world is limited to appearances which are always finite. Thus, the world as an infinite whole, as 'the world total', cannot appear to us. We should even say that such a world does not exist empirically for us at all.[10]

This notwithstanding, space and time as the subjective forms of intuition are infinite according to Kant: 'to say that time is infinite means nothing more than that any determinate magnitude of time is possible only through limitations [put] on a single underlying time. Hence the original presentation (*Vorstellung*) *time* must be given as unlimited'.[11] The same applies to space. It too may be added, as well as divided, infinitely.[12] However, two things, at least, should be kept in

---

6   KpV AA 5, 162.
7   See esp. KrV A 483/B 511. Kant writes: 'The absolute total of magnitude (the world total, das Weltall), of division, of origin, of the conditions of existence as such – with all the questions as to whether this total is to be brought about through finite synthesis or through a synthesis to be continued ad infinitum – concerns no possible experience.' For the experience 'the world total' is always comparative, never absolute.
8   Cf. KrV A19/B 33.
9   KrV B 72.
10 Cf. Josè A. Benardete, *Infinity: An Essay in Metaphysics* (Oxford 1964), 106.
11 KrV A 31–32/B 47–48
12 See Henry E. Allison, *Kant's Transcendental Idealism: An Interpretation and Defence* (New Haven 1983), 93–94.

mind here. Firstly, this notion of infinity belongs to those truths of transcendental aesthetics which we know a priori but never empirically. Secondly, Kant is using here a mathematical notion of infinity, i.e., one that is characterized by such negative terms as 'unlimited', 'endless', 'boundless', 'untraversable', 'unsurveyable', and 'immeasurable'.[13] 'The true (transcendental) concept of infinity is this: that the successive synthesis of unit(s) in measuring by means of a quantum can never be completed'.[14] As for Aristotle, mathematical infinity for Kant means basically a series which cannot be traversed or completed.[15]

Kant writes about a regress that cannot be completed.[16] We run into such a regress because the empirical information we receive about the world is always local and conditional. It is conditioned by the activities of our mind, first since time and space are the subjective forms of sensible intuition, and then by the categorical structure of our judgements. Everything that we experience as the world, and every particular object we know, is also conditioned by what we have before experienced and known. Our experiences always condition our empirical intuitions and, thus, initiate a regress that cannot be completed by any single intuition or experience. In this way we are led to a mathematical infinity, characterised by negative terms such 'unsurveyable'. Such an infinity belongs to our experiences, even if no single infinite intuition, or intuition of such infinity is given.[17]

Further, Kant maintains that both time and space, as subjective forms of intuition, are mathematically infinite. They are considered for conceptual reasons, for otherwise we could not manage the conditions of what we experience. That time and space are mathematically infinite is a synthetic a priori truth of the transcendental aesthetics.[18] Similarly, that infinity belongs to the categorical structure of understanding is a synthetic a priori truth of the transcendental analytic.[19] Thus in Kant's view, time and space as subjective forms of empirical intuition are infinite but everything that we intuit in space and time is finite. The infinite forms of intui-

---

13 On the distinction between mathematical and metaphysical infinity see A.W. Moore, *The Infinite* (London 1990), on Kant especially 84–95, as well as the essay 'Aspects of the Infinite in Kant', *Mind* 97 (1988), 205–223 by the same author.
14 KrV A 432/B 460.
15 On Aristotle's notion, see Benardete, *Infinity*, 1–18; Moore, *The Infinite*, 34–44; and Norman Kretzmann (ed.), *Infinity and Continuity in Ancient and Medieval Thought* (London 1982).
16 E.g. KrV A 432/B 460, A 527/B 555.
17 See the next section of this essay.
18 See esp. KrV A 30–32/B 46–48.
19 According to Kant, in a transcendental logic the quality of judgements is divided into affirmative, negative and infinite. Whereas in general logic infinite judgements are included in affirmative ones, here a judgement like 'The soul is nonmortal' 'must not be omitted from the transcendetal table of all moments of the thought occuring in judgements, because the function that the understanding performs in these infinite judgements may perhaps be important in the realm of the understanding's pure apriori cognition' (KrV A 72/B 98).

tion, as well as the categories of understanding, are given to us a priori, whereas everything that is given in them is a posteriori.

Kant admits that this idea is contrary to both common sense and to the tendencies or claims characteristic of human reason, because they both ask for the totality of conditions in a realist sense. Traditional metaphysics has answered to such demands by constructing theories about infinite totalities. These demands and their metaphysical answers are interpreted by Kant in the transcendental dialectic. He insists that although we are inclined to think of the world in space and time as infinite, this inclination, or need, should not be taken as a matter of knowledge. Rather, it is a matter of rational faith by which we orient ourselves in thought.[20] For Kant, we are dealing here with 'mere ideas' or ideals of reason. The idea of an infinite world may be highly important, especially for our practical use of reason and more generally for our rational orientation in life. Such ideas are not constitutive, however, but merely regulative ones, i.e., even if we can never know it, we may proceed as if the world were an infinite whole. From this perspective Kant discusses various traditional arguments pro and con concerning the infinite magnitude or divisibility of the physical world, organising them into the first two antinomies.

3.

Kant is not so much interested in the doctrinal history of cosmological views but rather in the indirect proof for his own transcendental idealism. His theory of time and space is tested particularly in the first two antinomies. Kant's strategy is to argue that if it turns out that no definitive view can be achieved for or against any claims concerning the infinity of the world in space and time, then we must conclude that the problem itself is wrongly posed and should be rejected. The antinomies arise because 'reason demands this totality according to the principle that *if the conditioned is given, then the entire sum of conditions and hence the absolutely unconditioned* (through which alone the conditioned was possible) *is also given*'.[21] Kant suggests that reason is unavoidably led to the four antinomies.[22] These are, therefore, an indirect test for his own transcendental idealism, especially for his theory of time and space as the subjective forms of intuition, as well as for his view on the different aspects of infinity.[23]

---

20 See Immanuel Kant, *What is Orientation in Thinking?*, in: Hans Reiss (ed.), *Political Writings* (Cambridge 1990), 237–249.
21 KrV B 436.
22 KrV B 490.
23 See Lothar Kreimendahl, 'Die Antinomie der reinen Vernunft, 1. Und 2. Abschnitt'; Eric Watkins, 'The Antinomy of Pure Reason, Sections 3–8'; and Henry Allison, 'The Antinomy of Pure

The *first antinomy* is approached by the question whether the world is infinitely old or large. The *thesis* states that 'the world has a beginning in time and is also enclosed within bounds as regards space'.[24] This is so in respect to time because every moment in the past belongs to a series of successive states, which thus cannot be without a beginning. In respect to space this is the case because, given that the space has no bounds, it is impossible to create a synthesis or even think of such a whole. The *antithesis* states that 'the world has no beginning and no bounds, but is infinite as regards both time and space.' This is true because both time and space as such are infinite and homogenous, so that no finite entity may exist as a boundary beyond which there is neither time nor space.[25]

In the *second antinomy*, the *thesis* states that 'every composite substance in the world consists of simple parts, and nothing at all exists but the simple or what is composed of it.'[26] This is the case because if the parts were infinitely divisible, it would follow that there are no parts at all. The *antithesis* suggests that 'no composite thing in the world consists of simple parts, and there exists in the world nothing simple at all.' This position is suggested, because every composite thing as well as its parts are in space, which is infinitely divisible. Thus, no thing may be simply finite.

For Kant, the thesis in both antinomies represents a Newtonian position and the antithesis a Leibnizian one, even if he calls the first 'dogmatic' and the second 'empiricistic'.[27] The fact that in neither case is there a definitive argument either for the thesis or its opposite is for Kant an indirect proof of his own view, i.e., that there is no such thing as the physical world as a whole in the sense that we can know by experience whether it is infinitely large and divisible. Kant makes use of sceptical argumentation as he concludes that the views of traditional cosmology are wrong because they exceed the limits of our possible knowledge. Instead of a metaphysical finitude or infinitude of the world we should, according to Kant, reflect philosophically the subjective structure through which the world appears to us. A mathematical infinity necessarily is built into the structure of our mind, but from this we should make no metaphysical inferences concerning the infinity of things in themselves.

Within this mathematical concept of infinity, in the discussion of the first antinomy, Kant makes further distinctions. 'The true transcendental concept of

---

Reason. Section 9', all in Georg Mohr, Marcus Willaschek (eds.), Immanuel Kant. *Kritik der reinen Vernunft* (Berlin 1998), 413–490 and the literature given there.

24   KrV B 454
25   On the alternative interpretation of details see Kreimendahl, 'Die Antinomie der reinen Vernunft,' 424–428 and the literature mentioned there. On the various criticisms of Kant's arguments with a defence of Kant, see Allison, *Kant's Transcendental Idealism*, 35–50.
26   KrV B 463.
27   Moore, *The Infinite*, 92.

infinity is this: that the successive synthesis of unit[s] in measuring by means of a quantum can never be completed'.[28] From this he distinguishes 'a defective concept of infinity of a given magnitude', i.e. the fact that there is such thing as the greatest number.[29] In the footnote to the above formulation about the transcendental concept of infinity, Kant further adds: 'The quantum thereby contains a multitude (of [a] given unit) that is greater than any number – which is the mathematical concept of the infinite.' Thus we have three notions of infinity, all of which are mathematical in the sense of my overall distinction between mathematical and metaphysical infinities. In the question concerning how we should understand the further distinction between transcendental and mathematical infinity, it is reasonable to understand the latter as a schematised version of the former.[30]

4.

In the passage about 'the starry heavens above me and the moral law within me', Kant uses the notion of infinity in its metaphysical sense. The starry heavens give rise to a truly sublime experience, undoubtedly an impressive and important one. Still more important, however, is the metaphysical infinity associated with the moral law within us. Of this Kant writes that it 'begins from my invisible self, my personality, and presents me in a world which has true infinity but which can be discovered only by the understanding'. This infinity is disclosed because, as he writes a few lines later, 'the moral law reveals me a life independent of animality and even of the whole sensible world, at least so far as this may be inferred from the purposive determination not restricted to the conditions and boundaries of this life but reaching into the infinite'.[31] Actually, this metaphysical infinity does not belong to us but to God, whose intuition is intellectual and creative, being thus not restricted by the empirical synthesis of the manifold.[32] As rational, moral persons who reflect motives and reasons for actions, however, we may participate in such a true infinity.

---

28 KrV A432/B 460.
29 KrV A430/B 458.
30 I follow here Allison 1983, who writes about this latter mathematical infinity: 'It contains a specific reference to number, the schema of quantity (KrV A 142/B182), and it expresses in numerical terms what the "transcendental", or "pure", concept presses in numerical terms; namely, the thought of the imcompletability or inexhaustability of the enumerative processes' (Allison, *Kant's Transcendental Idealism*, 42). – Allison writes also (42): 'Thus, whatever fault we may find in Kant's argument, I do not think that we can locate the trouble in his conception of the infinite.'
31 KpV 270.
32 See KrV B 72, 138–139, 145–146.

The purely rational will as such is unconstrained and metaphysically infinite. Although the true infinity of moral freedom cannot be known by experience, we may recognise its demands by means of a kind of moral intuition. Kant maintains that we may be aware of the moral law as our own free creation while sensing its demands on us. This infinity appears in the realm of experience, too, as we formulate moral maxims for our actions and judge them from the universal perspective of the categorical imperative, but it is always mixed with and conditioned by our fundamental finitude.[33] 'Act in accordance with maxims that can at the same time have their object themselves as universal laws of nature'.[34] The idea is that we should test our finite maxims from a perspective which is universal and thus infinite.

The metaphysical infinity of moral freedom is present for us in the form of ideas of practical reason. As moral persons we have, necessarily according to Kant, a moral faith in God as well as in the eternal life of our soul. 'No one, indeed, will be able to boast that he *knows* that there is a God and that there is future life (...) No, the conviction is not a logical one but a *moral* certainty; and because it rests on the subjective bases (of the moral attitude), I must not even say, It is morally certain, etc., but must say, I *am* morally certain, etc.'.[35] Even if we cannot have knowledge based on experience of it, the metaphysical infinitude of moral freedom demands faith in God as well as an infinite perspective in which we may think of finally becoming what we ought to be, i.e. worthy of our happiness. The postulates of practical reason are regulative ideas, i.e. we must act as if we knew that God exists and that we live eternally. In this 'as if', as we formulate and judge our maxims, the metaphysical infinity of moral freedom affects us as finite beings living in the realm of causal necessity.

Kant also establishes a connection between sublimity and the idea of the soul's immortality. Earlier in his precritical writings he associated God's sublimity with ideas about an immortal body-soul, writing about the sublimity of our 'coming eternity'. As he later, in his critical writings, gave up such metaphysical doctrines, he began instead to connect sublimity with our dual character as mortal and immortal beings. Interestingly, in his Third Critique Kant emphasises more strongly than before, our own finitude as revealed in the experiences of the sublime.[36]

According to Kant's analytic, an experience of the sublime is loaded with strong tension. It is caused by something that is either infinitely large or powerful and which, thus, threatens us. In the end, however, our reason manages to take this

---

33 See, e.g. KrV A 802–803/B 830–31.
34 Immanuel Kant, *Groundwork of Metaphysics of Morals,* in: Mary J. Gregor (ed.), *Practical Philosophy* (Cambridge 1996), 86 (AA 4: 437).
35 KrV A 829/B 857.
36 See the instructive essay by Olaf Briese, 'Ethik der Endlichkeit: Zum Verweisungscharakter des Erhabenen bei Kant,' *Kant-Studien,* 87 (1996), 325–347.

threat under control. These two moments are brought together in an experience of something pointing towards the infinite, or supersensible 'that nature in space and time [i.e. phenomenal nature] entirely lacks', namely the supersensible idea of nature in itself. 'But when intuiting nature we expand our empirical power of presentation (mathematically or dynamically), then reason, the ability to [think] an independent and absolute totality, never fails to step in and arouse the mind to an effort, although a futile one, to make the presentation of the senses adequate to this [idea of] of totality. This effort, as well as the feeling that the imagination [as it synthetisizes empirical nature] is unable to attain to that idea, is itself an exhibition of the subjective purposiveness of our mind, in the use of our imagination, for the mind's supersensible vocation'.[37]

Unlike in his early writings, Kant emphasises in the *Kritik der Urteiskraft* that 'true sublimity must be sought only in the mind [*Gemüt*] of the judging subject, not in the natural object, the judging of which prompts this mental attunement'.[38] The failure or limitation of aesthetic imagination brings about the experience of sublimity, as the imagination realises that its inability is ameliorated by reason. Only reason as a supersensible faculty may step over the limits into which the aesthetic imagination is confined. Reason exercises its power in such an experience, and it does so in a way which, according to Kant, points to its moral dimensions. Thus, Kant writes, 'if we judge aesthetically the good that is intellectual and intrinsically purposive (the moral good), we must present it not so much as beautiful but rather as sublime, so that it will arouse more a feeling of respect (which disdains charm) than one of love and familiar affection'.[39] Although an experience of the beautiful may symbolise morality and be psychologically favourable in moral education, only the sublime can bring us aesthetically into contact with the infinity of the moral law which demands on us respect.

Unlike beautiful forms, which give rise to harmonious and pleasurable experience, the sublime is associated with formlessness.[40] Whereas an experience of the beautiful contains a harmonious co-operation of imagination and reflective understanding, an experience of the sublime is both an agreement and a conflict of imagination and reason. Imagination encounters its limits and joins reason, which demonstrates its superiority. In the mathematically sublime, the superiority of reason is shown in the evaluation of magnitudes; in the dynamically sublime, it is

---

37 KU 268.
38 KU 256.
39 KU 271. On the connections between sublimity and morality see Paul Guyer, *Kant and the Experience of Freedom* (Cambridge 1993), 187–274; Wilhelm Vossenjuhl, 'Schönheit als Symbol der Sittlichen,' *Philosophisches Jahrbuch* 1/1992.
40 See KU § 23, 30, and Louis Roy, 'Kant's Reflection on the Sublime and the Infinite,' *Kant-Studien*, 88 (1997), 44–59, 45.

shown in the strength of the human mind to confront the forces of nature.[41] When confronting something that is infinitely large or powerful, we may 'discover in ourselves an ability to resist which is of a quite different kind, and which gives us the courage [to believe] that we could be a match for nature's seeming omnipotence. (...) a different and nonsensible standard that has this infinity itself under it; (...) an ability to judge ourselves independent of nature, and reveals in us a superiority over nature that is the basis of a self-preservation quite different in kind from the one that can be assailed and endangered by nature outside us'.[42] According to Kant's construction, the mathematically sublime reveals the intellectual superiority of human reason, whereas the dynamically sublime unfolds the superiority of moral reason.

In the sections on the sublime, Kant uses a distinction between two kinds of infinity which is operative already in his First Critique.[43] He writes about quantitative and aesthetic estimation of magnitudes and, related to this, about the quantitative infinite as 'an endless progression' and about the aesthetic infinite as a feeling of the sublime.[44] Corresponding to this, he writes concerning the first antinomy in the First Critique that 'the true (transcendental) concept of infinity is this: that the successive synthesis of unit[s] in measuring by means of quantum can never be completed'.[45] In the transcendental aesthetics, however, he writes about 'space as an infinite given magnitude' (...) whose 'original presentation' 'is an a priori intuition, not a concept'.[46] Thus there are two different kinds of a priori presentations of that infinite space which I have called mathematically infinite, an aesthetic intuition and a quantitative concept.[47] The same is true of time.[48] These two kinds of presentations of the mathematically infinite are complementary, which is also indicated by the discussion in the Third Critique. It seems plausible to think that the aesthetic presentation of the infinite is a prerequisite for the quantitative one.[49]

---

41   See KU § 27, 28.
42   KU 221.
43   Cf., Roy, 'Kant's Reflection on the Sublime and the Infinite', 54–59 and Jacob Rogozinski, 'Le don du monde,' in: Jean-Francoise Courtine (ed.), *Du sublime* (Paris 1988), 200–204.
44   See KU §26, 27.
45   KrV A432/B 460.
46   KrV A 25/B 40.
47   The distinction between mathematical and dynamical antinomies is of the same root. The two first, mathematical, antinomies are concerned with appearances qua intuited, whereas the latter two, dynamical, antinomies are concerned with conditions of experience that are thought rather than intuited.
48   See KrV A 32/B 48.
49   Cf., Roy, 'Kant's Reflection on the Sublime and the Infinite,' 56, who writes: 'The aesthetic estimation of greatness requires the idea of an absolute whole, which must be given by reason. This aesthetic estimation is presupposed by the mathematical estimation.'

5.

For Hegel, unlike Kant, reality (*Wirklichkeit*) is not behind appearances but manifests itself in the spatio-temporal world itself. Hegel follows Aristotle and in particular Spinoza in rejecting the dualism between reality and appearances. Kant was right, Hegel maintains, in defending the metaphysical infinity of reason, but he was fundamentally wrong in limiting reason to our capacity to practice freedom in the form of moral obligation. For Hegel reason is the principle of everything that is real and as such metaphysically infinite. Reason is not merely the moral principle; it is as well the foundational ontological and epistemological principle as well. Thus, when Hegel equates the real and the rational[50], this speculative identity should be read as a critical and rational perspective on what there is. From this perspective, the spatio-temporal world and its phenomena may be grasped as truly infinite and organised rationally.

According to Hegel's holistic rationalism, reality is a Spirit which first externalises itself into nature and then returns back to itself, returning into itself in the forms of consciousness and self-consciousness through the phenomena of human society and culture, most completely in art, religion and philosophy. In these forms of the Absolute Spirit the unity of reality and reason, i.e., the metaphysical infinity of everything, is grasped. For human life and its realisations are apprehended as a whole within an absolute perspective of all perspectives. All of the central notions of Hegel's metaphysics have the same holistic and infinite character: idea, subjectivity, substance, spirit, truth, freedom. They are all organised into syllogistic formations, pointing dialectically – in Hegel's sense – towards the rational unity of everything.

For Hegel, the true infinity is not the opposite of finitude but completes or contains this notion. Infinity is not a negative but a positive notion, although it contains a negation as its organising principle. Infinity is nothing separate from its finite parts, or elements, and correspondingly, knowledge of the infinite is nothing separate from the knowledge of finite phenomena. For Hegel, finite phenomena as something independent of the metaphysical infinity are 'positivities' which he intended to destroy. Hegel became committed to this program when he was in Frankfurt (1798–1800) in close relation with his friends Hölderlin and Schelling, striving to relate finitude to infinity within the metaphysical notions of love and life. During his first years at the university of Jena, he worked intensively on the logical and metaphysical problems of presenting and justifying such an infinite unity.[51]

---

50 See Hegel, *Elements of the Philosophy of Right* (Cambridge 1991), 20.
51 See Hegel, *Frühe Schriften. Werke I*. (Frankfurt 1986). An interesting interpretation of Hegel's early development concerning his logical and metaphysical notions is Thomas M. Schmidt, *Anerkennung und absolute Religion: Formierung der Gesellschaftstheorie und Genese der spekula-*

From his first philosophical publications, Hegel was convinced that we should and can cognise the truly infinite through the finite, and that the 'need of philosophy' is associated precisely with this cognition.[52] Thus, for Hegel Kant was wrong in thinking that external reality cannot be known as such and as a whole. Similarly, Kant and such reflective philosophers as Fichte and Reinhold, were unable to grasp true infinity, as they in various forms reproduced an opposition between the finite and the infinite. Hegel makes all this point clear in *Glauben und Wissen*, as he criticises Kant, Fichte, and Jacobi for being unable to grasp infinity in its true metaphysical meaning as a moment of the absolute. All three philosophers reproduce in various forms the fundamental opposition between the finite and the infinite, and Hegel's central claim is that they make this opposition – which is built into the modern culture as the opposition between faith and knowledge – into the principle of philosophy itself.[53] For them, the known world is finite, and in this finite world is felt 'the infinite longing that yearns beyond body and world, reconciled itself with existence'.[54] It is analogous to religious longing, presented in the form of a philosophical system.

Thus 'the fixed principle of this system of culture is that the finite is in and for itself, that it is absolute, and is the sole reality'.[55] Hegel claims that this principle of modern culture is transformed into a philosophical one: 'the fundamental principle common to the philosophies of Kant, Jacobi and Fichte is, then, the absoluteness of finitude and, resulting from it, the absolute antithesis of finitude and infinity, reality and ideality, the sensuous and the supersensuous, and the beyondness of what is truly real and absolute'.[56] They all reproduce such oppositions without a unity. Hegel's own conviction is that such an unity is possible when infinity, in its true meaning, is spelled out, and his critical reading of the Kantian, Fichtean and Jacobian systems again and again presses the points which aim at thinking such infinite unity.

From this general perspective Hegel briefly comments on the Kantian antinomies. Kant is unable, Hegel maintains, to solve the antinomies because 'Reason, as the dimensionless activity, as pure concept of infinitude is held fast in its op-

---

tiven Religionsphilosophie in Hegels Frühschriften (Stuttgart 1997). My critical review is published in: *Association* 3 (1999), 317–324. The best account of the development of Hegel's relation to Kant is Martin Bondeli, *Der Kantianismus des jungen Hegels: Die Kant-Aneignung und Kant-Überwindung Hegels auf seinem Weg zum philosophischen System* (Hamburg 1997). My review is published in: *Bulletin of the Hegel Society of Great Britain* 37/38 (1998), 112–117.
52 This was Hegel's main point in his first publication in Jena, *The Difference between Fichte's and Schelling's System of Philosophy* (Albany 1987).
53 See GW 55–57/317–317.
54 GW 317/58.
55 GW 319/60.
56 GW 321/62.

position to the finite. Reason is an absolute, and hence a pure identity without intuition and in itself empty'.[57] From the Kantian viewpoint, the antinomic views on the divisibility and magnitude of the world are not based on reason but on the transgression of the limits of human understanding. The conflict 'originates only through and within finitude and is therefore a necessary illusion.' Yet, Hegel comments, 'he did not succeed in dissolving the conflict. He did not succeed, in the first place, because he did not suspend the finitude itself. On the contrary, by turning the conflict into something subjective again, he allowed it to subsist. In the second place, he did not succeed because he can only use transcendental idealism as a negative key for the solution of the antinomy inasmuch as he denies that either side of it is anything in itself. In this way, what is positive in these antinomies, their middle, remains unrecognized'.[58]

It is obvious that Hegel's critical comments are built on premises which Kant himself would have rejected as dogmatic. The finitude of our empirical judgements can be suspended, or sublated, according to Hegel, when reason organises judgements into syllogisms and deduces through their middle terms new structures on higher ontological levels. From such a holistic and metaphysical viewpoint the Kantian solution, as sketched in the third antinomy between 'causality of freedom' and 'causality of necessity', i.e., the practical reason as the infinite progress, is not a solution at all, 'but confesses its finitude and its inability to validate its absoluteness precisely through this infinity of progress'.[59] At stake here are, more generally, two different conceptions of dialectics and of the role of contradiction in them. For Hegel, the important contradictions do not stem from mistakes or from the finitude of our understanding and reason but belong to reality itself and its rational cognition. Contradictions lead us through various finite phenomena towards the infinite, i.e., the absolute whole.[60]

In his critical discussion of Jacobi's philosophy Hegel explicates some of his notion of true infinity, which is much indebted to Spinoza. In his *Letters on Spinoza,* Jacobi accused Spinoza of a kind of slippery slope of infinite causal chains through which he attempted to explain the world without a transcendent God.[61] For Jacobi, this infinite application of the principle of sufficient reason leads necessarily to nihilism. Hegel defends Spinoza, and he wants to show that Jacobi has not understood Spinoza and especially his notion of infinitude. Jacobi has, he maintains, confused Spinoza's actual infinity (*infinitum actu*) with his

---

57 GW 336/81.
58 GW 337/84.
59 GW 337/84.
60 See Schmidt, *Anerkennung und absolute Religion,* 123–201 and my review; Bondeli, *Der Kantianismus des jungen Hegels,* 249–340.
61 See Frederick C. Beiser, *The Fate of Reason: German Philosophy from Kant to Fichte* (Cambridge, Mass. 1987), Ch. 2; Schmidt, *Anerkennung und absolute Religion.*

infinity of imagination. At stake is implicitely the disagreement between Christian theism and the more pantheistic vision of Spinoza and Hegel.

Spinoza defines the actual infinite (*Ethics,* Part I, Prop. VIII, Sch. I) as 'the absolute affirmation of existence of any nature'. Hegel's comments show that this view is close to his own view: 'Thus this simple definition makes the infinite into the absolute and true concept, equal to itself and indivisible, which of its essence includes the particular or finite in itself at the same time, and is unique and indivisible' (354/107). Important is that in this 'infinity of the intellect', the cognition of which is an 'intellectual intuition', nothing is determined or negated. No opposition between the finite and the infinite emerges. This is what Hegel throughout his career means by the true, genuine, metaphysical infinity, symbolised by a circle, or a circle of circles.[62]

On the other hand, Spinoza's infinity of imagination is, in Hegel's words, 'only reflection that posits and partially negates the finite; and this partially negated thing, which, when posited for itself and opposed to what is in itself not negated, to what is strictly affirmative, turns this infinite itself into something partially negated. The infinite, being thus brought into antithesis with the finite becomes an abstraction, the pure Reason, or the infinite of Kant'.[63] Hegel calls the latter also 'the empirical infinity' – later he uses terms like 'bad' or 'spurious' infinity[64] – for which Jacobi makes Spinoza responsible 'with no more ado, although no philosopher was ever farther removed from assuming anything of that sort'.[65] Because Jacobi himself is bound to the opposition between knowledge and faith, he is unable to conceive or approve Spinoza's idea of metaphysical or actual infinity.

Hegel further explains this idea by referring to Spinoza's example of 'the space enclosed by two circles that do not have a common centre' (...), 'the figure which he took as his authentic symbol'. In such a space, the inequalities possible are infinite, because 'it is the nature of the thing that surpasses any numerical determination'. 'There is in this bounded space an actual infinite, an infinitum actu. In this example we behold the infinite, which was defined above as absolute affirmation or absolute concept, set forth at the same time for intuition, and hence in the particular; here the absolute concept is the identity of opposites in actu. If these parts are kept apart and as parts posited as identical; if this particular qua particular is posited as actual and expressed in numbers, and if it is to be posited in its incommensurability according to the concept as identical, then the empirical infinity arises in the infinite series of mathematicians'.[66] Hegel's main

---

62 See esp. Hegel, *The Encyclopedia Logic* (Indianapolis 1991), § 95.
63 GW 355/108.
64 See Hegel, *The Encyclopedia Logic,* § 62, 63, 64.
65 GW 355/108.
66 GW 357–358/111–112.

theses in Glauben und Wissen is that in the modern culture, as philosophically conceptualised by Kant, Fichte and Jacobi, the actual infinite is present only as a matter of faith, as longing and striving, i.e., as a task. Spinoza is one of the very few who was clear about the actual infinite in its true metaphysical sense which should, Hegel insists, play 'a greater role in any exposition of the system than that which is alotted to it in Jacobi's propositions, where it is always just an idle predicate of thinking'.[67] Hegel's seriousness concerning this emphasis is indicated in the essay by the fact that he ends it with a sweeping recapitulation – he writes about a speculative Good Friday – filled with theological metaphors, of these two forms of infinity.

---

67 GW 358-113.

6. The Copenhagen Interpretation of Quantum Mechanics and the Question of Causality

Tarja Kallio-Tamminen (Helsinki)

A philosopher once said 'It is necessary for the very existence of science that the same conditions always produce the same results'. Well they do not. The characteristics of nature are not to be determined by pompous preconditions, but by nature herself.[1]

1. Introduction

In the beginning of the modern era, reality was sharply divided into two parts: *res extensa* and *res cogitans*. Physics began to study the measurable world of extended objects, whereas subjective phenomena belonged to a different realm. Newtonian mechanics presented a clear and visualizable model of the mechanical world where nothing could happen that was not in principle predictable. The behaviour of material particles was supposed to be governed by strict deterministic laws. The end of nineteenth century saw a common belief that physics had revealed the real nature of world. The power of the detached human mind had been able to catch an objective God's-Eye-View of the whole cosmos.

Visualizable Newtonian mechanics was replaced in the 1920's by the indeterministic and abstract formalism of quantum mechanics. Quantum mechanics entailed a radical shift of paradigm within physics. It offered the same kind of broad and universally applicable theoretical framework that was able to lead modern research as Newton's mechanics had done in its time. The interpretation of this irrevocably complex theory that deals with manydimensional spaces, however, has been difficult. In spite of the prolonged discussions a consensus concerning it is yet to be found. Even the basic question whether the theory has any profound implication for our understanding of reality and its metaphysical foundations has remained open.

Many problems emerged in connection with quantum mechanics that are difficult to conceptualize and understand within the visualizable framework of classical physics. For example, the interpretation of wave-particle dualism, the principle of uncertainty, and quantum non-locality have caused controversies for more

---

[1] Richard P. Feynman, *The Character of Physical Law* (London 1992), 147.

than 70 years. The founding fathers of the Copenhagen interpretation, who were also largely responsible for the construction of the theory, were convinced that quantum theory could not be satisfactorily interpreted and understood within the prevailing mechanistic and deterministic framework of classical physics. They proposed profound revisions to our ontological and epistemological approaches to reality. The ideas of Niels Bohr especially represented a radical reconsideration of the traditional metaphysics. The role of the human observer and his position in reality is understood in his framework of complementarity in a new way that reconstructs the strict dualism of Descartes.

Many of the post-Copenhagen interpretations of quantum mechanics such as David Bohm's causal interpretation or the many-worlds' interpretation aim to maintain the familiar metaphysical presuppositions of determinism, reductionism and the detached observer. When these interpretations have been considered within the framework of classical metaphysics, they have often been perceived as realistic. Comparaply the basically realistic tendency of the Copenhagen group is ignored, and their ideas are seen as either positivist or anti-realistic. The Copenhageans did not, however, consider themselves as positivists. They were explicitly trying to understand the nature of reality. The latter interpretations appear in light of their wider viewpoint as futile attempts to return to classical metaphysics by postulating unfounded auxiliary hypotheses. These post-Copenhagen interpretations can be seen in Kuhnian terms as efforts to hold onto a 'normal science,' i.e., classical metaphysics, even if new evidence might actually demand entirely new approaches.

2. The Indeterminism of Quantum Mechanics

The new features of quantum mechanics convinced the Copenhagen group that the theory enabled and even made necessary the rejection of some of the metaphysical preconditions of the previous physics. Illustrative for their starting point is the question of Max Born: How can we speak about objective reality when the atomic world cannot even reveal itself without the observer and his measurement arrangements?[2] The Copenhagen group came to see the objective world of nineteenth century science as 'a straightforward simplification, an ideal limiting case, that was not the whole reality'[3]. Classical physics presupposed that one could always in principle determine the state of a physical system to any desired degree of accuracy by measuring certain quantities. When a system of exceedingly small

---

2   Max Born, *Physik im Wandel meiner Zeit* (Braunschweig, Friedr. 1983), 59.
3   Werner Heisenberg, *Physics and Beyond* (London 1971), 88.

mass was considered in quantum mechanics, however, the state of the object was disturbed by the measuring process in an unpredictable way.[4]

The nature of this 'disturbance' was difficult to conceptualize. It was something inherent in nature, something that could not be overcome by better instruments or observation techniques. The interaction between the measuring instruments and the object under observation could no longer be neglected or totally controlled. Measurement was not possible without an unpredictable statistical impact produced by the measurement instruments; for this reason the state of the observed system was changed during the measurement process. In addition the principle of uncertainty removed the possibility of knowing precisely the mechanical state of the system under study. Even if in classical physics knowledge about exact position and momentum had been a prerequisite for determining the path of a particle as well as predicting its future, in quantum physics when the position of the studied system was determined precisely the experimenter was unable to determine the momentum of the system.

The Copenhageans believed that quantum mechanics had revealed a deeper underlying unity in nature. Niels Bohr spoke about the unsuspected limitation of the mechanical conception of nature as well as about the indivisibility of the measurement process.[5] He maintained that Planck's discovery of the elementary quantum of action had revealed a feature of wholeness inherent in atomic processes. Since the symbolic operators of quantum mechanics were liable to a noncommutative algorithm, the mechanistic and reductionistic space-time description of classical physics was prevented. Indeed, the usefulness of the mechanistic models within which all events could be described as independent objective movements in space-time was limited to the macroscopic world. In quantum theory we are beyond the reach of pictorial visualization. Furthermore, the demand of unrestricted divisibility, that all phenomena can be arbitrarily subdivided, is not applicaple. It is not even always possible to divide the subject from the object in an unambiguous way. The subject-object distinction is not fixed, and no clearcut distinction can be made between the objects themselves and their interacton with the measuring instruments. Observations can be obtained only under given experimental conditions.[6]

The Copenhagen group thus clearly recognized and drew serious attention to this novel holistic feature inherent in the world described by quantum mechanics. Bohr debated it for decades with Einstein, but only nowadays when the experiments of Alain Aspect have proved that the so called local realistic theories predict wrong results has the EPR-controversy has been settled in favour of Bohr. Recent

---

4   Arthur March, *Quantum Mechanics of Particles and Wave Fields* (New York 1951), 1.
5   Niels Bohr, *Atomic Physics and Human Knowledge* (New York 1958), 61, 74, 90, 92.
6   Niels Bohr, *Essays 1958/62 on Atomic Physics and Human Knowledge* (New York 1963), 2, 14, 92.

experiments on quantum coherences have also confirmed a certain entanglement of spatially separated systems. These results are still described as unexpected and counterintuitive[7] even if they would not have surprised the Copenhagen group. This implausible nonlocality is clearly anticipated by the formalism of quantum mechanics even if the results are difficult to understand within the accustomed framework of mechanically interacting individual particles.

Arthur March, a German professor of physics, tried to conceptualize the Copenhagen ideas on the foundations of quantum mechanics in a clear and careful manner. He reflected the new situation in his remark that any phenomenon that occurs in the micro-world consists of elementary processes or acts which, by virtue of natural law, cannot be analyzed. Hence, the micro-world is by nature atomic not only with respect to matter but also with respect to events. We shall never know what happens in an atom during the process that leads to the production or annihilation of a photon. The emission or absorption of light as well as the scattering of a photon by an electron are examples of elementary processes or acts which resist any attempt to analyze them. Therefore, we cannot apply the principle of causality to these processes. The atomicity of events appears to us as a discontinuity in the course of events, and only probability relations exist between present and future.[8]

The illusion of a predictable autonomous external world broke down on the micro level. Quantum mechanics did not allow for the possibility of making observations without reference to the observer or to the means of observation. Werner Heisenberg connected this problem to the indisputable fact that natural science bears the stamp of human beings. He supposed that the science of nature does not deal with nature itself but with of nature as man views and describes it. Natural science does not simply describe and explain nature as it is but is a part of the interplay between nature and ourselves. It describes nature as exposed to our method of questioning.[9]

3. *The Copenhagen Explanations for Indeterminism*

Quantum mechanics convinced the Copenhagen group that the new physics is essentially statistical. Thus, the world is not totally predictable for human beings and the classical concept of causality is inapplicable. Different situations produce different phenomena; human beings are not only passive observers but, by manipulat-

---

7   S. Haroche, 'Entanglement, Decoherence and the Quantum/Classical Boundary,' *Physics Today* (July 1998).
8   March, *Quantum Mechanics of Particles and Wave Fields,* 1–3, and Arthur March, *Das neue Denken der modernen Physik* (Hamburg 1957), 47–50.
9   Werner Heisenberg, *Das Naturbild der heutigen Physik* (Hamburg 1955), 18–21 (1962, 81.).

ing the boundary conditions, also design or shape the obtainable outcomes. According to Heisenberg, Descartes could not have attained such a concept. It makes the sharp separation between the world and the I impossible.[10]

The founding fathers of the Copenhagen interpretation clearly noticed that this critique of the very foundation of classical physics demands a new orientation. It is not easy, however, even today to acknowledge the consequences of this evidence. While a detached observer is not able to grasp all the mechanisms of material world, should we give up the idea of detachment or, perhaps better look for hidden variables that guarantee the strictly deterministic structure of the world even if we cannot ever detect them? Here is where an operationalistic physicist stops to speculate. She has no means to decide the question by experiment. Therefore, she contents herself with the idea that in quantum physics the concept of chance plays a role for which there is no room in the methods of classical physics. The idea of statistical causality is connected to those situations that cannot be exactly predicted.[11]

Even though the Copenhagen group admitted that in the observable world something is at work which can be designed only as chance, they were not content with just noticing it. They wanted to know why. Wolfgang Pauli and Werner Heisenberg suggested the concept of ontological renewal. Pauli did not believe that statistical causality was blind. He compensated the lack of natural cause by a metaphysical one by saying that nature makes a choice among the different possibilities that are allowed by the state function. Pauli even presented a speculative ontological framework that involved consciousness. In his psychophysical and irrational reality cosmic architypes were able to control and harmonize mind as well as matter. They chose the results they preferred, sometimes, perhaps, by working through the subconscious mind.

Heisenberg also directed his attention to the concept of *causa*. It had had a much broader meaning in earlier philosophy, but in the beginning of the modern era the universe was stripped of personal or supernatural aspects. The physical world was no longer a visible expression of spiritual reality but was opaquely material and possessed no deeper intrinsic meaning. Only the *causa efficiens* of the four causes asserted by Aristotle was adopted as the determining cause of every event in nature. Heisenberg did not hesitate to suggest that the new results of physics gave credibility to some Aristotelian ideas. He assumed that probability meant a tendency toward something. It was a quantitative version of the old concept of 'potentia' in Aristotelian philosophy. It introduced something standing in the mid-

---

10   Werner Heisenberg, *Physics and Philosophy* (New York 1962), 81.
11   March, *Quantum Mechanics of Particles and Wave Fields,* 9–11.

dle between the idea of an event and the actual event, a strange kind of physical reality lying between possibility and reality.[12]

Niels Bohr was also convinced that the nature of ultimate reality could no longer be taken for granted. He carefully considered the tacit metaphysics of classical physics and could not even accept the fundamental subject-object dualism. He laid stress on the fact that natural scientists acquire their knowledge by actively participating in the processes of reality. They do not know in advance what kind of reality they are exploring, so they are not able to start from any given ontology. This fact doesn't mean, however, that Bohr would have doubted the existence of the external world. He emphasized the interaction between human beings and the rest of the world. We are observers and *actors*, not just onlookers but also participators in the drama of existence. But because in quantum reality phenomena are necessarily interconnected, contextual, and indivisible and because the observer and his instrument are embedded in that reality, the behaviour of objects under study can never be captured as they are in themselves. Thus, the scientist is unable to totally separate himself from the object of his study or to take the viewpoint of an objective external observer, as was supposed within classical physics. Bohr drew serious attention to this epistemological lesson taught by quantum mechanics. He tied in his complementarity scientific description firmly to human experience; but, at the same time, he also struggled to create a new exhaustive viewpoint that would overcome the limits inherent in the previously supposed God's-Eye-View.

Bohr paid serious attention to the nature of language in the formation of knowledge. He particularly wanted to understand why we have to use seemingly incompatible concepts such as waves and particles to describe empirical observations concerning atomic systems even though no real paradoxes can exist, because the unequivocal mathematical formalism of quantum mechanics guarantees the consistency of quantum description. He came to the conclusion that it is no longer possible to presume that our language 'corresponds with reality' in the way that was postulated in classical physics. By proceeding into areas far beyond our ordinary experience and immediate observation, the new physics had shown that the scope of classical concepts previously held as universal was rather limited. Our classical concepts are adapted to the macroscopic world and they simply do not fit to the atomic world, which does not allow familiar partitions and space-time models.

Bohr distinguished clearly between mathematical symbolic schemes and intuitive modes of description yielded by classical concepts. He believed that the mathematical formalism of quantum mechanics revealed structures of reality that simply could not be captured by ordinary language. Pure mathematics can not,

---

12 Heisenberg, *Das Naturbildt der heutigen Physik,* 24; Heisenberg, *Physics and Philosophy,* 40.

however, evoke real understanding of the nature of reality. Understanding the content of the theory requires retaing to the macroscopic experience and familiar classical concepts and analogies. We are bound to the use of well-defined familiar images and idealizations from our daily experience, even though they are inferior to mathematics for coping with quantum reality.

Bohr challenged through his notion of complementarity the alleged objectivity of classical physics. His approach is not, however, subjectivistic. Scientific theories contain the best intersubjective understanding of reality at any present time, even though one can never know whether they can capture reality in its entirety. When quantum mechanics made it evident that our classical language and models can not correspond with ultimate reality, Bohr did not conclude that we should give them up. On the contrary, he looked for a new comprehensive standpoint concerning the world and our role in it. He noticed that we can communicate all information obtained from the atomic world in terms of our familiar classical concepts. We have to use complementary descriptions, however, in different situations. For example, in certain experiments a given atomic object must be described as a wave, whereas in others it must be treated as a particle. These complementary descriptions may seem contradictory, but they actually complete each other. The concept of waves and particles are applicable in different situations, and they offer varying perspectives to reality. They are not arbitrary constructions, because they do refer to real phenomena. Atomic objects, however do not need be limited to either a wave or a particle form, even if our experiences in the macro-world have provided us with only these two images. Since the data that we acquire in complementary experiments and descriptions also contains the impact of the instruments used, our observations do not necessarily reach to the independent existence of the micro-world but merely to phenomena that come forth in the interaction. In other words, the apparent conflicts within complementarity are not caused by paradoxes within reality itself but by inconsistencies of our own mechanical models and classical language.

Bohr emphasized that the principle of complementarity allows the use of classical causal representation in such macro-level situations that are independent from the act of observation. At the micro-level, however, this exception is not applicable and accordingly, our own participation cannot be avoided. Thus physical description has to be reduced to the wider framework of complementarity. It provides a rational generalization of the previous ideal of causality and certainly does not refer to any limitations concerning quantum mechanical description.[13]

---

13 Bohr, *Atomic Physics and Human Knowledge,* 82 and Bohr, *Essays 1958/62 on Atomic Physics and Human Knowledge,* 6. See further Tarja Kallio-Tamminen, *Observers and Actors: The Role of Human Beings within Classical and Quantum Physics.* University of Helsinki, Department of Philosophy, Doctoral thesis, forthcoming.

4. The Unsolved Problem of Measurement

The philosophical path of deep conceptual and metaphysical revision paved by the Copenhagen group has not nowdays been followed in physics. Rather, paved the basic presuppositions and objective ideals of classical metaphysics are taken for granted. Trust in abstract mathematics is often connected to instrumentalism, which naturally does not produce any further understanding concerning the nature of reality. The chosen road has lead to countless successes when quantum mechanics has been adapted to different sorts of problems. Physicists, however, have not been able to find a consensus concerning measurement.

Abner Shimony described the conceptual problem of the reduction of a superposition as a 'small cloud' in contemporary physical theory in which its laws are otherwise completely independent of the existence of minds. He anticipated that our present intellectual discomfort would be compensated if this difficulty eventually provides some insight into the mysterious coexistence and interaction of mind and matter. 'Small clouds' in an otherwise highly successful theory have often been precursors of great illumination.[14]

It is generally assumed that a 'complete' theory should be applicable to microscopic objects as well as experimental setups. In classical physics the process of measurement could be described objectively from outside by comparing the result given by measurement to a space-time picture or model created by the theory. Quantum theory, however, does not generate a model for the process of measurement. Thus, it is not able to describe how it is possible to proceed from the uncertain and unclassical quantum realm to the stable and separable world of everyday experience. When the theory is applied to measuring apparati as well as to microscopic objects the final state of the system will be a superposition of state vectors. It does not, however, represent a definite observable state. The final state should be a so called mixed state, but no transformations map initial pure states into final mixed states. Apparently, then, according to quantum mechanics, measurement should be impossible.[15]

Physicists have tried many ways to establish substantive criteria for classicality within a quantum framework in order to give a better treatment of classical characteristics. They include superselection rules, decoherence effects, systems with infinite degrees of freedom and generalized observables. Most distressingly, none of them seems satisfactory; and results lead to inconsistencies that cannot be resolved in an obvious way.[16] An increasing number of physicists have begun

---

14 A. Shimony, 'Role of observer in Quantum Theory,' *Ann. Jnl. of Physics*, vol 31, (1963), 773.
15 Dugald Murdoch, *Niels Bohr's Philosophy of Physics* (Cambridge 1989), 113.
16 Sunny Y. Auyang, *How Is Quantum Field Theory Possible* (New York 1995), 82.

to suspect that this open problem of measurement may have far-reaching consequences for the possibility of recognizing an objective reality in physics.[17]

While the universality of the theory is taken for granted, it is still difficult to understand why the objectification of measurement has not succeeded. Why has the 'complete' theory, contrary to the accustomed way of thinking, not been able to generate a model for measurement? Should we, perhaps, conclude that the solution may indeed demand a completely new starting point instead of mathematical manipulation. Maybe the traditional dualistic framework of thinking does not offer tools for solving the dilemma. As a matter of fact the question concerning the interpretation of quantum theory also culminates in the measurement problem. The conflict between Copenhagen and post-Copenhagen ways of approaching reality and positioning human beings in it can be illustrated in their different treatments of the measurement problem. While the post-Copenhagen interpretations have tried to retain classical ideas concerning measurement, some have suggested that the Copenhagen solution to the measurement problem is to say that there is no solution.[18] I would rather say, however, that for Bohr there is no problem.

As a matter of fact, the Copenhagen interpretation has dealt with measurements in two different ways. Both of them emphasize the role of the observer, but in totally different ways. In his orthodox theory of measurement John von Neuman, in 1932 tried to find an traditional objective solution to the problem by introducing a concept of the collapse of the wavefunction. This postulate states simply that measurement causes an abrupt and irreversible transformation in which a pure state turns into a mixture of states. The development of the quantum mechanical system could, thus, happen in two totally different ways. Measurements should be treated differently from all other interactions in nature. Von Neuman did not provide any clear explanation for his projection postulate; but later the concept has lead to idealistic and subjectivistic interpretations according to the suggestion that consciousness somehow collapses the wavefunction.

Actually there is no evidence that this collapse really takes place. It is commonly thought that speech about collapses, reductions or jumps can be only metaphorical by the standard of physics, difficult as it may be to avoid if one wants to treat measurements in the accustomed classical way. Niels Bohr's approach offers a constructive alternative. Bohr was silent on the question of the collapse of wavefunction. He did not accept the common requirement that quantum theory should generate a model for the process of measurement. Neither did he believe that the traditional treatment of the measurement problem was possible. For Bohr quantum mechanics demanded the rejection of the whole idea of the causal space-

---

17 Peter Mittelstaedt, *The Interpretation of Quantum Mechanics and the Measurement Process* (Cambridge 1998), ix, 103.
18 Jim Baggott, *The Meaning of Quantum Theory* (Oxford 1992), 194.

time description of classical physics.[19] When it is not possible to create a visualizable model from the micro-world, neither is it possible to give a true description of how to proceed from the micro-level to classical reality. Classical language is not able to give an exhaustive model of the measurement process.

This failure does not mean, however, that any real difficulties arise concerning measurement. Bohr avoided the whole problem by treating measurement apparati in classical terms. The macroscopic world and all the marks of measurement instruments can be described precisely in classical language. Nor does the quantum mechanical treatment of measurement create a problem for Bohr, because the superposition states of macroscopic systems cannot in practice be differentiated from mixed states. For Bohr, the purpose of measurement is not to check a value of some pre-existent quantity but rather to get information and knowledge about reality by participating in its processes. When the system under study is not unambiguously separable from the measurement apparatus during the measurement process, the system is not closed and, accordingly, the human impact on the observed reality does not create any physical problem. The course of so called objective reality is actually changed in the different processes following our divergent choices and actions. The choices concerning different kinds of measurement methods and apparati are part of our irreversible activities. In the framework of complementarity the idea of universal predestination shall be replaced by the concept of natural evolution.[20]

5. Conclusions

Niels Bohr's framework of complementarity offers a possibility of avoiding the measurement problem. It also removes all of the apparent paradoxes of quantum mechanics without need to add any auxiliary hypothesis to the theory. His approach would, however, entail radical changes to our understanding of reality. Criticizing the subject - object dualism challenges the accustomed foundations of natural sciences. Descartes proclaimed that neither God nor any rational soul present in World will ever disturb the ordinary course of nature in any way[21]. Saying 'no' to this foundational dualism would probably collapse the paradigm of the western conception of reality that has characterized the modern era. The post-Copenhagen interpretations try to rescue us from such a radical shift, being perhaps comparable to the transformations that took place in antiquity and at the

---

19 Henry J. Folse, *The Philosophy of Niels Bohr: The Framework of Complementarity* (New York 1985), 68–70.
20 Bohr, *Atomic Physics and Human Knowledge*, 81.
21 Rene Descartes, *Discourse on Method and Related Writings* (London 1999), 112.

turn of the Early Modern era when the most profound metaphysical presuppositions concerning reality were also changed.

Clinging to out-dated presuppositions, however, may prevent a proper understanding of the role of human beings in the world: what are the actual limits and possibilities of our existence? The dualism of Descartes gave birth to the mind-body problem, but it was not able to provide any elucidation or justification for it. Neither could the problem be satisfactorily solved by an idealism or materialism that could not overstep the traditional mechanical view concerning physical reality. By allowing indeterministic and irreversible processes, however, quantum mechanics is able to give a preferable approach to understanding man's relation to nature. In quantum reality our role and existence can be viewed from new perspectives; especially is it easier to locate our mental activities within the actual physical world.

Bohr in his approach replaced the passive observer of classical physics with an active participant who is deeply interconnected with reality. The knowledge, values, and goals of human beings directly influence the distribution of possibilities in reality through our choices and actions. As our knowledge about reality is expanding, we are increasingly able to influence even areas of nature that were previously thought to be objectively given or beyond human reach. Yet, the increase of influence is bound to increased responsibility. Quantum mechanics made it clear that we are not allowed to consider ourselves as totally omnipotent agents who are able to rule and manipulate the deterministic realm of matter as we like; for we are deeply influenced and dependent on the contextual processes of nature. Our actions do have an effect both on ourselves and on our future circumstances.

Bohr's viewpoint is compatible with the emerging understanding concerning symmetries and complex systems. Herman Weyl has already said in his book *Symmetry:* 'The truth as we see it today is this: The laws of nature do not determine uniquely the one world that actually exists.' The important laws of conservation in physics are governed by simple symmetrical principles. These symmetrical principles may lead to many asymmetrical outcomes; hence, as soon as a system has symmetry, a good chance arises that the symmetry may break. Symmetry breakage gives rise to the growth of complexity in the universe by generating outcomes that are more complicated than the laws themselves. When a symmetry breaks, very tiny asymmetries presumably play a crucial role in selecting the actual outcome from a range of potential results that are allowed by natural law.[22]

I think there is no point to deny that human beings as well are very much able to generate some of these symmetry breakages. Their meaning can be minor

---

22 Stewart and Gikubitsky, *Fearful symmetry* (London 1993), 15–17, 54.

on the scale of the universe, but the quality of our local neighborhood depends decisively on our own actions as well. They are not unaffected by our beliefs or by our theories. For the sake of our future I think it would be time to update our conception of reality. This magnificent complex system is not a simple clock-like world.

7. Subject and Object in the Study of Nature

Sami Pihlström (Helsinki)

1. Introduction

The radical changes in our traditional Western conceptions of reality that were apparently required by some of the well-known results of quantum mechanics, especially in its Copenhagen interpretation, have puzzled both physicists and philosophers for most of the 20th century. This article, written as a response to Tarja Kallio-Tamminen's paper,[1] is an attempt to look at the situation from a rather general philosophical point of view without a detailed discussion of the problems involved in the interpretation of quantum mechanics. In particular, I want to take up the notions of subjectivity and freedom. I shall argue that we should be very careful in drawing philosophical conclusions regarding these notions from any physical theory, classical or non-classical. Even though my explicit critical comments are restricted to Kallio-Tamminen's views, I intend them to be taken seriously by any philosophizing physicist or by any philosopher interested in the metaphysical significance of the world-view of contemporary physics.

I must confess right in the beginning that I find many of the more or less metaphorical formulations of the 'new paradigm' provided by the quantum-theoretical picture of the world, formulations often employed by 'Copenhagenian' philosophers working on the foundations of quantum mechanics, rather vague. This vagueness makes the philosophical evaluation of the Copenhagen interpretation difficult. For example, Kallio-Tamminen's remarks about 'the objective world' as a 'simplification' and an 'ideal limiting case', about 'deeper underlying unity in nature', and about 'undivided individuality' and 'wholeness' need precising. So also do her remarks about 'the classical concept of causality', about the equally 'classical' idea that our language should 'correspond with reality', about the 'traditional dualistic framework of thinking', and about our being 'deeply interconnected' with reality. All of these notions and many more ought to be made much more precise. I do not have in mind any scientific or technically philosophical or logical preciseness here. Sometimes perhaps just historical accuracy is what is required. For instance, what exactly do we mean by the classical concept of

---

[1] See Tarja Kallio-Tamminen, 'The Copenhagen Interpretation of Quantum Mechanics and the Question of Causality,' this volume. All unspecified quotations in the text are from Kallio-Tamminen's essay.

causality – a Humean, empiricist regularity theory or a non-Humean, realist view of causality as a genuine feature of the physical world? We must leave these matters of clarification, however, for other investigations. Presumably, similar demands for clarification can be presented with respect to the philosophical vocabulary I shall employ in the following.

2. Subjectivity and Objectivity

Despite her occasional conceptual impreciseness, Kallio-Tamminen's Kuhnian approach in the philosophy of physics is interesting and certainly worth taking seriously. Her thesis that the recent anti-Copenhagen interpretations of quantum mechanics (e.g., David Bohm's and his followers' ontological interpretation) are attempts to make sense of the puzzling features of quantum mechanics within a Kuhnian 'normal science' (i.e., the deterministic and realistic paradigm of physical science that has dominated Western thought since Galileo, Descartes, and Newton) should, of course, be tested further through historical and philosophical investigations. The 'new paradigm' of the Copenhagen interpretation might, if Kallio-Tamminen is right, still be emerging – although most physicists, I am afraid, may not agree with her claim. Someone might even urge that the Copenhagen paradigm itself has proved to be a blind alley and that other, less positivistic and more realistic approaches in the interpretation of quantum mechanics ought to be explored. Since I am not a physicist, it will be safer for me to leave this controversy untouched here.[2]

Independently of the problem of interpreting quantum mechanics, what I find difficult to believe or sometimes even to understand in Kallio-Tamminen's ideas regarding the (allegedly novel and revolutionary) world-view of the Copenhagen interpretation is her insistence, inspired by Niels Bohr, to conceptualize human observers and their relation to the world in an essentially 'non-dualistic manner'. My worries here are, I am tempted to say, inspired by a Kantian view. I believe that adopting a Kantian perspective is justified in this context, for the general issue concerning realism (whose special case is the realism issue as applied to non-classical physics and which thus forms the background of the interpretation problems of quantum mechanics) is largely a creation of post-Kantian philosophy.

Now, it is certainly plausible to argue that no sharp separation 'between the world and the I' is possible in the sense in which Descartes intended it. Such separation, and possibly even its rejection, remains, however, within what Immanuel

---

2 Hence, I shall not pretend to be able to judge whether the Copenhagen philosophy is accurate as an interpretation of quantum mechanics today. For an account of some of the more recent developments in quantum physics, see Kari Enqvist, 'Time and Causality in the Theory of Everything,' this volume.

Kant more than 200 years ago called 'transcendental realism'.³ The problem of whether or not 'the I', construed as an immaterial metaphysical entity somehow belonging to the world, exists independently of its natural or material surroundings is, from a Kantian perspective, the wrong place to begin investigating the relation between subjectivity and objectivity. Kant rejected Cartesian dualism and its assumption of metaphysically existing, immaterial, rational 'souls' as emphatically as any quantum theorist would; yet, he never doubted that our experiences have a basic subject-object structure. As subjects of experience we are 'active', just as the Bohrian subject of quantum-mechanical measurement is thought to be. Well known is that Kant tried to show that the subject necessarily imposes certain general features upon the experienceable world, such as spatiality, temporality, and causality. Objects can only be experienced as conforming to these subjectively structured 'epistemic conditions' of experience.⁴

If this view is correct, the very objectivity of the objects of the world requires that they are given in experience to an active, organizing, transcendental (but not transcendent or other-worldly) subject. Accordingly, no object – whether macroscopic or microscopic, classical or non-classical⁵ – can, from the Kantian perspective, be studied as it is in itself; to imagine such a study as possible would be to conflate empirical appearances with things in themselves. Such a conflation is the root of the difficulties of transcendental realism, which assumes the possibility of a transcendent, absolute 'God's-Eye-View' of the world and thus conceptualizes human knowledge according to a theocentric model.⁶ It is precisely the Kantian requirement of investigating the world as it appears to us, however, as a human world, that necessitates a transcendental (non-empirical) subject-object duality. In their somewhat overhasty rejection of any distinction between the world and 'the I', Bohrian philosophers of physics like Kallio-Tamminen still work, it seems to me, within the same pre-Kantian problem framework in which their opponents like Cartesian dualists operated.⁷

---

3   On Kant's distinction between transcendental idealism and transcendental realism, see Immanuel Kant, *Kritik der reinen Vernunft* (Hamburg 1990), A28/B44, A35–36/B52–53, A369ff., A491–497/B519–525.
4   The expression 'epistemic conditions' is taken from Henry E. Allison's famous interpretation of Kant. See his *Kant's Transcendental Idealism: An Interpretation and Defense* (New Haven 1983).
5   Kant, of course, knew nothing about non-classical physics. The obvious fact that his Newtonian conception of the physical world is outdated is, however, entirely irrelevant to the philosophical significance of his 'Copernican revolution' and to the general problem of realism to which his views can be seen as providing an answer.
6   See Allison, *Kant's Transcendental Idealism*, ch. 2.
7   By the way, it is somewhat odd that in her paper, Kallio-Tamminen only considers Descartes as the 'bad guy' whose doctrines are to be given up (only?) after the rise of quantum mechanics. Why does she not discuss Kant at all and the clear similarities (and differences) between

The subject-object structure that Kant insisted upon is required, in particular, by the very idea that we (as Copenhagenians following Bohr, are often inclined to say) are agents taking part in the processes of the world; that is, we are not only observers of nature but also active participants 'in the drama of existence'. No such participation is possible without some distinction between the one who (actively) participates and the system (or world) in which she or he participates. I believe this is a rather simple and valid transcendental argument that locates a necessary condition for the possibility of any activity or agency in human subjectivity: as subjects we are not reducible to mere objects of nature. Nor should we forget that several philosophers have already discussed the idea of our participation in the drama of existence before the emergence of the quantum theory.[8] Thus, we hardly need the new physics in order to make this view philosophically interesting or defensible. On the contrary, if we are led to abandon all distinctions between the subject and the object, the notion of participation is rendered obscure.

The role of subjectivity, however, should not be overemphasized, or, rather, it should not be emphasized in an incorrect way. We should, in particular, be suspicious of idealistic and potentially pseudo-scientific accounts of how human consciousness 'collapses' the wave-packet. Such theories can be compared to the kind of 'empirical idealism' that Kant found in Berkeley and rightly rejected.[9] No Kantian philosopher should believe that empirical objects and events (classical or quantum-physical) are mere products of the human mind. What Kant's *transcendental idealism* is all about is the necessary subject-imposed (transcendental)

Kantianism and the Copenhagen interpretation? It seems to me that the Copenhagen interpretation is strongly Kantian in its rejection of the possibility and even meaningfulness of a 'God's-Eye-View' to the world; on the other hand, it is quite clearly non-Kantian in its insistence that philosophical problems of subjectivity and freedom can be solved by studying, quantum-physically, the empirical world. (I shall return to this issue below.) On the relation between Kantianism and the Copenhagen philosophy, see, e.g., Hilary Putnam, *Realism with a Human Face* ed. by James Conant (Cambridge, Mass. 1990), ch. 1.

8   The American pragmatist William James was one of these thinkers. See especially his late work *A Pluralistic Universe* ed. by Frederick H. Burkhardt, Fredson Bowers and Ignas K. Skrupskelis (Cambridge, Mass. 1977). On James's pragmatism as a form of ontological constructivism emphasizing our active participation in the structuring of reality, see, e.g., Sami Pihlström, *Structuring the World: The Issue of Realism and the Nature of Ontological Problems in Classical and Contemporary Pragmatism* (Helsinki 1996); *Pragmatism and Philosophical Anthropology: Understanding Our Human Life in a Human World* (New York 1998).

9   Allison's above-mentioned book (1983) provides a useful account of why Kant's transcendental idealism must not be confused with Berkeleyan phenomenalism. See also Henry E. Allison, *Idealism and Freedom: Essays on Kant's Theoretical and Practical Philosophy* (Cambridge 1996); David Carr, *The Paradox of Subjectivity: The Self in the Transcendental Tradition* (Oxford 1999); Arthur W. Collins, *Possible Experience: Understanding Kant's Critique of Pure Reason* (Berkeley 1999).

conditions for the possibility of obtaining empirical knowledge of objective, mind-independent things. Kantian appearances are real, not imagined or illusory. The transcendental activities of the subject should not be identified with any mystical theory of a 'consciousness' which is somehow active on the quantum level. Such theories can only lead to what Kant regarded as transcendental illusions.

Let me note at this point that I am in sympathy with Kallio-Tamminen's advocacy of Bohr's view which refuses to see any genuine problem in the collapse of the wave function. Kallio-Tamminen sometimes seems, however, to take a rather idealistic turn herself when she begins to speak about our choices and actions changing the 'objective' world. If this were intended as a transcendental thesis about certain subjective epistemic conditions being constitutive of the objectivity of things existing in the empirical world, it would be quite acceptable from a Kantian point of view. I am afraid, however, that it is intended as a (quasi-)empirical thesis about the interaction of subjective choices and worldly events, a thesis to be arrived at on the basis of the quantum theory. As such, it is hard to believe, and, even if it were true, it would assume the subject-object duality that was supposed to be given up in the Copenhagenian philosophy. Saying (with Kallio-Tamminen) that our 'knowledge, values, and actions directly influence the distribution of possibilities of reality' is almost a commonplace in a Kantian transcendental setting. As an allegedly scientific or science-based claim about the relation between the knowledge, values and actions of empirical human subjects in the empirical world described by this interpretation of quantum theory it sounds magical rather than philosophically interesting.

The upshot of our Kantian analysis can be extended to the methodological assumptions at work in the advocacy of the Copenhagen interpretation. The Kuhnian conception of scientific research as something that a social group sharing a paradigm engages in (a conception to which Tarja Kallio-Tamminen appeals in her defense of the revolutionary character of the Copenhagen standpoint) requires the same basic subject-object structure as Kant's transcendental idealism and is in that sense inescapably Kantian. It is the paradigm, or the scientific community working within a paradigm, that is the subject of scientific experience; the paradigm can only change in interaction with its object, nature, i.e., something that it (in a certain sense, transcendentally) constitutes or 'constructs'.[10] We should, I think, be prepared to reinterpret Kuhn's philosophy of science in this Kantian manner, replacing Kant's rigid categories of understanding by the more flexible social structures established and critically revised within one or another scientific paradigm that 'structures' or 'shapes' the world for the relevant scientific com-

---

10 See Thomas S. Kuhn, *The Structure of Scientific Revolutions* (Chicago ²1970). For some comparisons between Kuhn and Kant, see Paul Hoyningen-Huene, *Reconstructing Scientific Revolutions: Thomas S. Kuhn's Philosophy of Science,* (Chicago 1993). On Kuhn's philosophy as a form of ontological constructivism, see Pihlström, *Structuring the World,* ch. 4.4.

munity. At any rate, if we employ Kuhn's basic position in our analysis of the development of modern physics, we should take this Kantian approach seriously.

More generally, from a Kantian or Kuhnian perspective, we hardly need the quantum theory at all to tell us that natural science itself is a human enterprise dealing with 'the interplay between nature and ourselves'. Endorsing this active and humanistic conception of the scientific enterprise, we should resist the temptation to fall into tautologies such that 'the science of nature does not deal with nature itself but with the science of nature as man thinks and describes it'. This view seems to be either harmless and trivial (and thus acceptable to any hard-boiled scientific realist or anti-Copenhagen quantum theorist) or just false. Clearly, science deals with nature as we think of it and describe it. Yet, it can hardly be the task of physics to simply inquire into how we think or have thought about the natural world. Such a task belongs primarily to the history of physics and, perhaps, to the philosophy of physics. What physicists are doing, and what most of them no doubt suppose they are doing, is to investigate the natural world itself. It is another matter how the natural world and its physical describability and explainability are to be philosophically interpreted. I tend to agree with Copenhagenians like Kallio-Tamminen that the classical idea of the absolute objectivity of the world is problematic and that something much more modest than a metaphysical realism committed to a deep (perhaps ultimately unknowable) truth about the way the world in itself is will be needed.[11] I doubt, however, that we need quantum theory in order to take a step toward that kind of a weaker realism. What we need is *philosophical* criticism of our habitually realistic and objectivistic ways of thinking. Such criticism has been going on in academic philosophy at least since Kant, who was one of the most important classical authors to question our almost instinctive uncritical belief in the absolute objectivity of the (empirical) world. He reminded us that what we experience or cognize is inevitably a world of appearances whose objective structuredness is grounded in the constitutional activity of the transcendental subject, yielding the (transcendental) epistemic conditions on the basis of which (alone) objective knowledge is possible.

In sum, I fail to see how a Copenhagenian quantum theorist like Kallio-Tamminen can consistently entertain either her metaphysical concept of human beings as active participants in the world's scheme of things or her methodological and

---

11 Hilary Putnam has for many years insisted that metaphysical realism is closely linked with the idea of truth as non-epistemic correspondence between statements and the mind- and language-independent world. His own 'internal realism' rejects this not just as false but as absurd, arguing (pragmatistically) that truth cannot be separated from human interests and values. See Putnam, *Realism with a Human Face*, as well as, more recently, Hilary Putnam, *Words and Life* (Cambridge, Mass. 1994). For a critical discussion of Putnam's recent views on realism and truth, see Pihlström, *Pragmatism and Philosophical Anthropology*, ch. 3. It is an open question whether Kallio-Tamminen or other Copenhagenians would be willing to subscribe to a Putnamean epistemic and pragmatic conception of truth.

historical conception of the present situation in quantum mechanics, analyzed in terms of Kuhnian paradigms, unless she is willing to soften the Bohrian abandonment of the subject-object dualism. Finally, let me briefly note, without further argument at this point, that the Kantian concept of 'the I' as a transcendental subject actively constituting the world of experience (phenomena, appearances) is by no means dualistic in Descartes's metaphysical sense. The transcendental subject is not a Cartesian 'rational soul' at all. Such souls, construed as immaterial (quasi-) objects in the world as seen from a 'God's-Eye-View', were rightly dismissed already in Kant's first *Critique,* quite independently of quantum mechanics. Thus, in order to dispense with Cartesian dualism, we do not need modern physics; we just need Kant (and perhaps some post-Kantian thought). The Kantian transcendental subject is, contrary to the metaphysical subjects existing in a Cartesian universe, a formal subject that makes experience and the experienced world itself possible. Such a subject is, a Kantian might argue, required even as a transcendental ground of quantum-mechanical experience. It is hard to see, however, how any account of subjectivity based on developments in the sciences could adequately account for this transcendental subjectivity. If our participation in the world is explored on the basis of quantum mechanics, we can hardly transcend the empirical subject, whose inseparability from what takes place in 'objective' nature need not be denied but whose 'other aspect', transcendental subjectivity, ought as well to be acknowledged.[12]

3. *Free Agency*

Let me conclude with a couple of critical comments on freedom that also reveal my Kantian doubts concerning the viability of the overall project of some Copenhagenian philosophizing physicists (like Kallio-Tamminen's). When our examination of human subjectivity is conducted on a transcendental (instead of empirical or scientific) level, we should become extremely doubtful about the relevance of any scientific theory to such properties of the subject as freedom. I can hardly understand the claims that quantum mechanics would somehow save 'freedom of the will' through its proofs of indeterminacy.

Tarja Kallio-Tamminen, among others, appears to assume such freedom in her account of how we can change the course of the objective world through our acts. Again, a Kantian *compatibilist* solution to the problem of freedom is in my view

---

12 Carr's above-mentioned work, *The Paradox of Subjectivity,* rightly reminds us that the transcendental and the empirical subject can be regarded as two aspects of one and the same entity. There is no need to postulate two separate realms of either objects or subjects in the Kantian picture. Cf. here also Sami Pihlström, 'Investigating the Transcendental Tradition,' *Philosophy Today* 44 (4) 2000, 426–411.

more sophisticated: we are not free as empirical subjects, as denizens of the objective natural world (whether deterministically or indeterministically conceptualized); but we should nonetheless think of ourselves (that is, of the very same beings who, empirically considered, are not free) as being free when considering ourselves as moral agents, and, hence, responsible for our actions. This move views human subjectivity in two different though complementary rather than simply incompatible ways.[13] The relation between this Kantian complementarity of two conflicting but equally necessary descriptions and its modern Bohrian counterpart remains to be investigated. This is not the right place for such an investigation, but we should note that there are clear analogies to be found here: just as the 'wave picture' and the 'particle picture' of the microphysical level of the world conflict with each other, though both are needed for a 'complete' quantum-physical description, so we cannot avoid conceptualizing our lives subjectively as series of actions conducted by free and responsible agents, while also asserting our best scientific accounts of the blind and meaningless determination of natural events. The challenge is to view the 'same' events as both meaningless (i.e., merely natural or factual) and meaningful in a specific human way (i.e., as free actions). I am not saying that this would be easy to do; what I am saying is that we are dealing here with an irreducibly philosophical problem, a problem not to be solved by reference to empirical science.[14]

---

13 This doctrine of 'two standpoints' is to be distinguished from a metaphysical notion of two different worlds inhabited by human beings; see again Carr, *The Paradox of Subjectivity*. According to Carr, the transcendental tradition has always recognized that we are paradoxical creatures in that we are at once both subjects and objects. It seems to me that those (usually Copenhagenian) interpretations of modern physics that try to make room for freedom in the physical world rather superficially hide the paradoxical nature of our existence. No scientific picture of the world, quantum-theoretical or classical, can alleviate the tension between our being, on the one hand, empirical objects interacting with other parts of the natural world and, on the other hand, free and morally responsible subjects. On Kant's compatibilism (i.e., the view that the freedom of the will and the causal necessitation of whatever takes place in the empirical world are compatible), see also Hud Hudson, *Kant's Compatibilism* (Ithaca, N.Y. 1994).
14 There is another equally fundamental problem confounding any attempt to suggest that a philosophical puzzle can be settled through scientific theorizing. How do we know that the results of quantum theory, or any other theory, are sufficiently 'final' in the sense that we have (say, with the Copenhagen interpretation) arrived at a super-paradigm faithfully revealing us something that is 'inherent in nature' (i.e., indeterminacy and lack of rigid causality)? Shouldn't we, especially if we take seriously Kuhn's account of the development of science (as Kallio-Tamminen does), be more modest and admit that new conceptual revolutions may occur even within fundamental physics, yielding novel, unpredictable paradigms? Were we capable of such modesty, we would perhaps be less eager to draw deep philosophical lessons from the current state of scientific theory. In my view, philosophers should be extremely careful about appealing to any scientific theory as the guide to a rational world-view. (This is not to say that philosophers should not be interested in the results of science. It is just to say that those results cannot dictate the outcome of our philosophical arguments, although they may influence it.)

Another way of putting the matter is as follows. Freedom, instead of being scientifically defensible (or criticizable), is, as G.H. von Wright (among others) has argued for decades,[15] *conceptually* related to the notion of agency. That is, we cannot make sense of our ineliminable experiences of agency without employing the notion of freedom; and this is so irrespective of whether that notion can be accommodated by some natural-scientific theory. No scientific theory, and *a fortiori* no interpretation of the quantum theory, can achieve this intrinsic conceptual link between the concepts of freedom and agency nor investigate any philosophical problems involved in that relation. Instead of trying to 'solve' the problem of freedom by claiming that quantum mechanics somehow secures a role for it in the microlevel structure of nature, we should perhaps simply admit, with von Wright, that '[t]o deny that an *agent* is *free* is to commit a contradiction in terms' and that, accordingly, the 'mystery' of our freedom, 'if there is one, is the "mystery" of the fact that there *are* agents and actions'.[16] This mystery is something quite different from any scientific account of the microstructure of our world. To claim that there are 'agents' or 'actions' involved in whatever it is that the quantum theory describes and explains is, I am inclined to think, to commit a category mistake.

Recognizing the reality of our problem of reconciling these two apparently conflicting perspectives on human life, i.e., the subjective and the objective perspective (or the perspective of agency and freedom, on the one hand, and the perspective of the natural (in)determinacy of whatever takes place in the natural world), is to be willing to live with uncertainties and tensions in one's world-view. Such a willingness is, I think, truly philosophical. From the Kantian point of view that I have very briefly outlined here, there are good reasons to oppose the various projects for overcoming subject-object dualism that are designed to take away the inevitable complementarity we should find in our existence as both subjects and objects.[17] The picture I am recommending, however, may be inherently unstable. While I have tried to defend the idea that neither the subjective nor the objective conception of humanity should have primacy over the other (since both are needed in a humanly adequate self-understanding), I must admit that any comparison,

---

15 See especially Georg Henrik von Wright, *Explanation and Understanding* (London 1971); *Freedom and Determination* (Helsinki 1980). For a recent discussion of this aspect of von Wright's thought, with a clarifying recognition of his association with the Kantian tradition in this respect, see Rosaria Egidi, 'von Wright and 'Dante's Dream': Stages in a Philosophical Pilgrim's Progress,' in: Rosaria Egidi (ed.), *In In Search of a New Humanism: The Philosophy of Georg Henrik von Wright* (Dordrecht 1999), 1–34.
16 von Wright, *Freedom and Determination*, 77–78 (also quoted by Egidi, 'von Wright and 'Dante's Dream', 16).
17 Not only the Copenhagen interpretation of quantum mechanics (which some critics take to be idealistic), but physicalistic and materialistic accounts of human beings as simply parts of nature as well, are such projects of unification, in my view equally vulnerable to a Kantian critique.

assessment, or choice between these conceptions will inevitably be made from the subjective, agent-centered (and thus non-scientific) point of view in which freedom and responsibility are already assumed to be actual. One can engage in a Kantian (or von-Wrightian) comparison between the different aspects of our paradoxical subjectivity only *as a free agent,* as a subject living not just in a physical, scientifically describable world but in a normative universe of discourse. Such engagements, furthermore, cannot be accounted for by any scientific theory, even though developments in natural science (e.g., the rise of quantum mechanics) may surely lead us to look at our philosophical problems in a new way.[18]

Far from pretending to be able to solve the problem of the two perspectives, I would, finally, be prepared to argue (again on a Kantian basis) that freedom, instead of being quantum-theoretically explainable, is primarily an *ethical* rather than a natural-scientific notion and that all attempts to defend freedom and ethics (or, analogously, religious views) in a natural-scientific way (on the basis of quantum mechanics or anything else) are doomed to end up with irresolvable confusions. This is, in effect, to repeat the point made above with reference to von Wright: freedom is structurally related to human agency, and agency is always, inevitably, ethically problematic. Isn't ethics (as well as religion) and, hence, freedom in the end something far more important than scientific descriptions and explanations of the world? How *could* ethics ever be defended (or refuted) by scientific means; how could it even be in the need of scientific support? To ask this question is not only to reject the specific metaphysical picture of freedom that some Copenhagenian interpreters of the quantum theory take to be justified; it is to put in doubt the basic motivation of such philosophizing based on physics.[19]

---

18 For some reflections on this double-aspect view of human nature, see Pihlström, *Structuring the World,* ch. 5.3.
19 I am grateful to Tarja Kallio-Tamminen for her comments on an earlier draft. I also wish to thank Dr. Rudolf Larenz for interesting conversations related to the topic of this paper.

8. Causes, Causes, Causes

Three Aspects of the Idea of Cause

Jaakko Hintikka (Boston)

1. *Prologue*

One cannot hope to do justice to the idea of cause in one paper. Important insights have nevertheless been reached into this notion recently which can be surveyed briefly here. In the following, I will present three episodes from the life of this important concept. By so doing, I hope to throw light on the idea of cause and its history as well as to provide an example of what philosophical analysis of crucial concepts like cause should be like.

2. *Aristotle on Causes*

There is no philosopher in the history of human thought whose views on causation have been more influential than Aristotle. But what were his views? Any competent graduate student will answer: Aristotle considered causation within the framework of a syllogistically organized science. Hence what is caused must be expressible by the minor term of a syllogism, for instance C in

(1) every B is A
every C is B
ergo: every C is A

In our contemporary notation this can be written

(2) $(\forall x)(B(x) \supset A(x))$
$(\forall x)(C(x) \supset B(x))$
ergo: $(\forall x)(C(x) \supset A(x))$

Alternatively, what is caused according to Aristotle is what is expressed by the conclusion ('every C is A') of a syllogism like (1) or (2). In either perspective, the cause is the middle term B. Any competent philosopher can tell you this. But can you tell me why Aristotle held these views? It is safe to say that in the voluminous literature on Aristotle you cannot find an informative answer to this question.

I will offer such an answer. In preparation to it, I note first that for Aristotle the ideas of cause and explanation are practically identical. His main term aitia can be translated, and has been translated, both as *cause* and as *explanation*. This close connection between the two notions is not a peculiarity of Aristotle's. For us, too, explaining why something happened and identifying its cause are almost indistinguishable. I will return to this matter later.

Unfortunately, this does not by itself explain Aristotle's views. We can make some progress by considering the quest of Aristotelian causes as an attempt to explain the conclusion of a syllogism, say 'every C is A', in symbols

(3) $\quad (\forall x)(C(x) \supset A(x))$

Explaining (3) involves deriving it from some explanatory premise or premises that tell us something about C and A. Now the syllogistic terms were for Aristotle not merely verbal formulas. They expressed universals or forms which were independent actual ingredients or factors in the universe. It may not be clear what this means or might mean, but one thing is clear: Since the different forms are different entities, we can in principle obtain knowledge of them independently of each other. Hence explaining (3) can be thought of as a derivation of it from two different bodies of knowledge, K[A] and K[C], the one about A and the other about C. Since Aristotle did not distinguish logical (conceptual) necessity from natural necessity, we can think of the derivation as a deductive one:

(4) $\quad (K[A]\ \&\ K[C]) \vdash (\forall x)(C(x) \supset A(x))$

so that

(5) $\quad (K[C]\ \&\ C(x)) \vdash (K[A] \supset A(x))$

Then by Craig's interpolation lemma there is an interpolation formula B[x] not containing C or A such that

(6) $\quad (K[C]\ \&\ C(x)) \vdash B[x]$

(7) $\quad B[x] \vdash (K[A] \supset A(x))$

But (6) can be rewritten

(8) $\quad K[C] \vdash (\forall x)(C(x) \supset B[x])$

and (7) as

(9) $K[A] \vdash (\forall x)(B[x] \supset A(x))$

What this means is that the explanation might as well be thought of as syllogistic, with B[x] as the middle term. Hence Aristotle's general metaphysics of forms or universals explains why explanations had to be syllogistic for him.

But why is the middle term the cause? An explanation lies in the independently ascertainable way in which Aristotle handled existence in a syllogistic context. Aristotle did not think of the different Frege-Russell senses of being (identity, predication, and existence) as different meanings of *estin*. Rather, they were different components of the unequivocal meaning of being which on different occasions could be present or absent from its force. When the other component senses were absent, *estin* expressed existence. However, existence pure and simple could not serve alone as a syllogistic term (predicate) of a scientific premise, not because this was logically impossible, but because it was too wide a term to be accommodated by any one departmental science. None the less, a syllogistic term could have or not have existential force, and that existential force could be conveyed from one term to another, as e.g. in the following syllogism:

(10)   every B is (an existing) A
       every C is B
ergo: every C is (an existing) A

Thus in a chain of scientific syllogisms existence is conveyed from higher (more general) terms to narrower ones. What conveys existence to the minor term C in (10) is therefore the middle term B of the atomic (i.e. minimal) syllogism of this form. According to Aristotle, C therefore exists because of B. And this is scarcely distinguishable from saying that B is the proximate cause of C, just as Aristotle maintained. By the same token, the ultimate cause is given by the most general term of a science, that is by the generic premise of this science.

Aristotle's theory of causation is thus seen to be a virtually unavoidable consequence of his general metaphysical and logical views. One consequence is therefore that if Aristotle's metaphysics of forms is given up, as the nominalists of the middle ages were the first ones to do, his notion of cause loses its systematic interest.

3. *Cause in Quantum Theory*

The second episode in the career of the notion of cause takes place much later. It is often said that in quantum physics the classical notions of cause and determinism lose their meaning. Let me try to analyze what this might mean.

In other papers I have offered an outline of what the logic of quantum theory is. The crucial idea is that in quantum theory certain pairs of variables, called noncommuting variables, are mutually dependent. This does not seem to be much of a novelty until one asks: How are dependencies and independencies between variables expressed in a logical language? The answer is obvious: By the dependence and independence of the quantifiers to which they are bound. And how is that kind of dependence and independence expressed? Everybody who has taken an introductory logic course knows the answer: By means of the parentheses that indicate the scope of a quantifier. (This makes parentheses the most important logical symbols.) For instance, in

(11) $(\forall x)((\exists y)(S[x,y]))$

the quantifier $(\exists y)$ depends on $(\forall x)$ and accordingly the variable y depends on x. (In (11) I have for the sake of clarity included the two pairs of parentheses which are usually suppressed.) This is indicated by the (dispensable) parentheses in (11).

But the use of parentheses as scope indicators and indeed the entire notion of scope is now seen to be inadequate. For *mutual* dependence cannot be expressed by their means. A quantifier $(Q_1 y)$ cannot be in the scope of another one, say $(Q_2 x)$, and at the same time vice versa. Scopes of quantifiers are nested, hence dependencies between quantifiers can be expressed only if they are transitive and asymmetrical.

What is needed is a more flexible logic, which is precisely what I have constructed in cooperation with Gabriel Sandu. It has been called independence-friendly logic, even though this term has turned out to be less than completely happy. By its means or, more fully expressed, by its means and on the basis of the working hypothesis mentioned earlier, viz. that noncommuting variables are mutually dependent, it can be argued that several of the main conceptual problems of quantum theory can be understood. Important cases in point are the measurement problem and nonlocality problem. This is strong evidence to the effect that mutual dependence of noncommuting variables is indeed a hallmark of the quantum logic.

But what does this have to do with the notion of cause? But which notion of cause? The classical one? Now whatever the classical Laplacean notion of cause is or may be, one unmistakable thing about it is that it is asymmetrical. The relation of a cause to an effect is different from the relation of the effect to its cause. Hence if the characteristic dependencies in quantum theory are symmetrical, they cannot be captured by means of the classical asymmetrical concept of cause, which is therefore inadequate for the purposes of quantum physics.

In the light of such results, it should not be a surprise to anyone that there has turned out to be dependencies in quantum phenomena that are not causal in the usual sense, for instance nonlocal dependencies.

This is one sense in which it is true to say that the classical notions of cause has been overthrown in quantum physics. But it is important to note that this does not mean that the notion of law-governed dependence has to be given up. On the contrary, the reason why the received notion is inadequate is that it does not apply in more complex patterns of dependence and independence.

4. *Causes and Explanations*

What, then, remains that we could reasonably call cause? Here the close connection between the ideas of cause and explanation turns out to be handy. Once again I can only summarize a long and complex line of thought.

A new theory of explanation has recently been developed by Ilpo Halonen and myself. The main idea is simplicity itself. To explain something according to basic idea of this theory is simply to derive the explanandum E logically from a suitable background theory T plus a number of explanatory data A, for instance from such observations and experiments as yield boundary conditions or initial conditions for the explanation in question. It suffices for my present purposes to consider a simple explanandum of the form P(b) as a test case.

Now this framework might seem to be far too simple to offer any insights into explanations, let alone causes. However, if we can make one innocent-looking further assumption, the situation changes dramatically. Let us assume merely that the predicate P occurs only in T = T[P] but not in A[b], and vice versa for b. Then we can consider the following consequence relations (12)-(14), all of which are equivalent with the given one (12),

(12) $(T[P] \ \& \ A[b]) \vdash P(b)$

(13) $A[b] \vdash (T[P] \supset P(b))$

(14) $T[P] \vdash (A[b] \supset P(b))$

If we apply certain simple logical results to (13)-(14) we obtain expressions H[x] and L[P] such that

(15) $A[b] \vdash H[b]$

(16) $T[P] \vdash (\forall x)(H[x] \supset P(x))$

(17) $T[P] \vdash L[P]$

(18) $L[P] \vdash (\forall x)(A[x] \supset P(x))$

These results are obtained from a strengthened version of Craig's interpolation theorem. The sentence A[b] can be understood as telling what it is about the explanatory situation (independently of T[P]) that necessitates the applicability of the 'covering law'

(19) $(\forall x)(H[x] \supset P(x))$

in the particular explanatory situation in question. Thus H[b] (and/or the general law (19)) explains why P(b) should be the case by spelling out what it is about the data A[b] that necessitates (together with the background theory T[P]) the explanandum.

Likewise, L[P] tells what it is about the theory T[P] that forces the explanandum to be true if the data are true. It might be called a 'local law'. Both are naturally thought of as explanations of the explanandum, although not in the same sense, and in suitable circumstances either one can naturally be called the cause of the explanandum.

If so, the notion of cause is systematically ambiguous. Moreover, there are even further senses of cause. It lies close at hand to try to identify the proximate cause of the explanandum by locating the last nonlogical step in the explanatory argument leading from T[P] and A[b] to P(b). This idea needs further development, however. For instance, the order of different steps in the derivation (12) may be varied, yielding different *prima facie* proximate causes for the same effect. Furthermore, the derivation (12) may fall into several alternative paths, only one of which will be actualized. The actual cause will then be specified by the last nonlogical step in the relevant derivation branch.

Furthermore, there will have to be some additional constraints as to when an explanation can be said to express a cause. Apart from all those details, however, it is eminently clear on the basis of such an analysis that the notion of cause is at best multiply ambiguous, and hence not very useful as an explanatory concept. Once again, the idea of dependence turns out to be more basic than that of a cause, for the theory of explanation structured above can be thought of as being a theory of explanation as dependence analysis.

5. *Causes in Everyday Life and in the Law*

There is one more way of extending this analysis so as to throw light on the way we actually use the notion of cause. In scientific contexts, it is usually fairly clear what the background theory T is. However, in principle the identification T is a pragmatic rather than logical or epistemological matter. This freedom can be used for the purpose of understanding the use of the notion of cause in everyday or legal contexts. In such contexts, we are frequently interested in the contribution of some particular factor – an agent, a policy, the weather, or whatnot – to what happened. One reason might be to allocate praise and blame, thus honoring the etymological connection between words for cause and guilt. In such a case, what we might want to do is to include all our knowledge of the other factors to T, i.e. to what in purely scientific contexts corresponds to the background theory. Then we could apply the theory of explanation and find out what it is about the designated factor that gave rise to the explanandum. For instance, if there has been a train crash, the investigative board might very well ask what it was about the train engineer that led to the disaster. In so doing, the board will have to consider the other factors as being given, such as the train company's policies and procedures, the weather, the condition of the track, the location of signals, etc., even if from another perspective they could very well have contributed to the actual cause of events. Once we have located the explanatory feature of the engineer, we can in normal circumstances happily (or unhappily) say that it was the cause in question. For instance, we could say that a train accident was caused by the engineer's failure to notice a signal (proximate cause) or by his ambition to keep the train on time that led him to accelerate the train (local law).

Without trying to offer further evidence, it seems to me that we can in this way capture the logic and epistemology of a large number of interesting uses of the notion of cause. These uses are pragmatically determined, but they all share the same logical structure and can therefore be analyzed in the same way.

6. *A Methodological Epilogue*

Even though an analysis of these three episodes does not amount to a theory of causation, they throw sharp critical light on the nature and history of the notion of cause. The analyses I have essayed have also a methodological moral. What I have done is something that is not only rare in these days but is in the mind of many philosophers (and non-philosophers) politically incorrect. I have relied on logical analysis as the most important tool of my analyses. (It is especially interesting to see that the logical results needed to understand Aristotle are the same as are needed to understand our contemporary problem of explanation.) Thus reliance on logic rather than on sociology makes a big difference, not only to the fashionableness or lack thereof of my work. Someone might try to align my analyses and those of my French namesake and say that my analyses are nothing but so many exercises in deconstruction. In fact, I do not necessarily mind this term. However, this kind of systematic logical analysis has the great advantage over Derrida in that it reveals the true structure underlying the contextually determined surface phenomena and hence yields the means of reconstructing the deconstructed ideas. On second thought, perhaps I do mind being associated with Derrida. He is giving deconstruction a bad name. What his writings illustrate is the old truth: Too little logic is a very dangerous thing.[1]

---

1   The following texts are considered relevant for this paper: I. Halonen and J. Hintikka, 'Aristotelian explanations,' *Studies in the History and Philosophy of Science* 31 (2000), 125–131; I. Halonen and J. Hintikka, (forthcoming), 'Toward a Theory of the Process of Explanation'; J. Hintikka, 'On Aristotle's notion of existence,' *Review of Metaphysics* 52 (1999), 779–805; J. Hintikka (forthcoming), 'A Logic for Quantum Theory.'; W. Salmon *Four Decades of Scientific Explanation* (Minneapolis 1990).

9. Quantum Mechanics and Emergence*

I.A. Kieseppä (Helsinki)

1. Introduction

In *The Mind and Its Place in Nature*, C. D. Broad claimed that although scientists might be able to deduce from their theories a large number of properties of ammonia, they would never be able to deduce from their theories *what ammonia smells like*. According to him, this inability showed that having the smell of ammonia is an *emergent property* of ammonia.[1]

More precisely, C. D. Broad defined the property of 'a whole' to be emergent if it could not be deduced from 'the most complete knowledge' of the properties of the parts of 'the whole'.[2] It is clear why the smell of ammonia seems to be an emergent property according to this definition. Just like colors and sounds, smells are *secondary qualities,* and reductionist explanations of secondary qualities seem impossible in principle: it is, indeed, difficult even to form a clear idea of what a reductionist explanation of a secondary quality is. How would one explain that, let's say, red things look red? One can explain how electromagnetic radiation affects human eyes, and how these effects are transmitted in the human brain, but it seems that such explanations can never include the fact that radiation of a particular wave length produces the particular *subjective experiences of redness* that we have.

As a matter of fact, writing shortly before the discovery of quantum mechanics, Broad doubted whether one could deduce even *the ordinary chemical properties* of ammonia from the properties of its 'parts', i.e. from the properties of the nitrogen atoms and hydrogen atoms of which ammonia molecules consist. In our time chemists have become able to deduce with the help of quantum mechanics many properties of molecules from the properties of the atoms of which they are made. This capacity, however, does not seem to affect the validity of Broad's main point, i.e. that reductionist explanations of secondary qualities seem impossible in principle.

Of course, secondary qualities are not the only examples of emergent properties whose status is problematic. For example, quantum mechanics and the micro-

---

* I am grateful to Kari Enqvist for our discussions concerning the subject matter of this paper.
1 C. D. Broad, *The Mind and its Place in Nature* (London ⁵1949), 71.
2 Broad, *The Mind and its Place in Nature*, 61.

physical theories which are based on its principles provide us with a mathematical and highly abstract description of reality, but contemporary *cosmological* theories discuss things like galaxies and stars, whose nature seems to be much easier to understand. Nevertheless, all contemporary cosmological theories presuppose that quantum mechanics, or some other theory which is based on its principles, is the correct theory of the universe on the microphysical level. This makes sense only if e.g. the property of being a star and the property of being a galaxy are *emergent* properties of sufficiently complicated quantum mechanical systems of some specific kind.

Within the restricted context of the philosophy of the mind another important and enigmatic class of emergent phenomena is constituted by the *propositional attitudes*, i.e. states of the mind such as believing or knowing. In every-day language a person might characterize her mental state by saying that she knows that p or that she believes that p, where p is a sentence, and it seems obvious enough that a satisfactory theoretical account of mental phenomena must take into account the phenomena which these every-day language expressions try to characterize. I.e., it seems obvious that a satisfactory account of mental phenomena must include an explanation why people have beliefs and knowledge. Yet it is hard to see how a scientific theory which deals with brain processes could ever end up describing anything which resembles knowledge or beliefs; it seems obvious that such a theory is bound to discuss only physical entities such as molecules and cells which a neurologist can actually observe in the brain.

The position of eliminative materialism asserts that the neuroscience of the future will make no reference to beliefs or other similar intentional mental states. According to this extreme view, our common-sense conception of psychological phenomena is fundamentally mistaken, and beliefs and other similar states of mind simply do not really exist.[3] In the context of the mind-body problem the doctrine of emergence means that, opposed to what eliminative materialists claim, mental phenomena such as states of belief do exist, and are *properties of matter which has been organized in a suitable way.* When molecules are organized to form cells, and when the cells are organized to form a human brain, their totality can have new features such as thinking that p, which individual molecules cannot have.

More generally, *emergentism* can be characterized as the doctrine that sufficiently complicated systems can have novel and emergent properties, such as e.g. that of being a galaxy, which their parts, such as e.g. individual molecules, do not have. But what does one mean, precisely speaking, by a novel or an *emergent* property? Different answers to this question lead either to stronger or weaker

---

3   See e.g. P. M. Churchland, 'Eliminative Materialism and the Propositional Attitudes,' *The Journal of Philosophy* 78 (1981), 67–90.

forms of emergentism. We saw above that for Broad the essential feature of an emergent property is that it *cannot be deduced* from the properties of the parts of the system.[4] More recently, emergent properties have also been taken to be the properties of a whole which can have *causal force* over its parts.[5] For example, the state of believing that a book X is interesting is an emergent property of a person in this sense of the word 'emergence' if it can cause her to borrow the book from a library and if this act cannot be explained by appealing to the laws that govern the individual cells and molecules in her brain.

One can also find in the relevant literature definitions of an emergent property whose conditions are weaker. According to one of these, a property of a system is emergent when a detailed knowledge of the properties *of the parts of the system* does not suffice to yield knowledge concerning the emergent property in question, since a knowledge of the *interrelations* of such parts is required as well.[6] Mario Bunge has presented a definition whose conditions are weaker still. According to Bunge, a property P of a system is emergent if no component of the system possesses P.[7] The doctrine of emergent properties is, of course, trivially true if the notion of emergence is understood in the sense of either of these definitions.

Kari Enqvist (in this volume) characterizes the relation between a macroscopic, coarse-grained descriptions of physical reality and its microphysical description as a relation of *weak emergence*. It seems that the notion of emergence that Enqvist has in mind is similar with that of Bunge, since he states that the coarse-grained description *can be deduced* from the underlying reality, although it is in some sense irreducible to it. As a similar example of weak emergence, Enqvist quotes the relation between the temperature of a gas and the movements of its molecules.

Clearly, this relation is a relation of emergence in the sense of Bunge's definition, but not according to the other definitions that we have considered. The tem-

---

4 More precisely, with this definition one wants to express the idea that an emergent property is not deducible from the combination of sentences which express properties of the parts of the considered system and the theories which describe (physical or other) laws of nature which govern the system. This more precise characterization seems to imply that the notion of emergence can be defined only relative to the set of the theories that one may use in the relevant deductions. This conclusion has been accepted in e.g. C.G. Hempel and P. Oppenheim, 'Studies in the Logic of Explanation,' *Philosophy of Science* 15 (1948), 135–175, 151. Clearly, if the notion of emergence is defined only relative to the existing theories, a previously emergent property can turn into a non-emergent one when a theory which explains it is discovered (cf. Hempel & Oppenheim, 'Studies in the Logic of Explanation').
5 The notion of emergence is used in this way in e.g. R. W. Sperry, 'Mind-Brain Interaction: Mentalism, Yes; Dualism, No,' *Neuroscience* 5 (1980), 195–206, in particular 201–2.
6 See e.g. P. Teller, 'A Contemporary Look at Emergence,' in: A. Beckermann, H. Flohr, and J. Kim (eds.), *Emergence or Reduction? Essays on the Prospects of Nonreductive Physicalism* (Berlin 1992), 139–153, 141.
7 M. Bunge, 'Emergence and the Mind,' *Neuroscience* 2 (1977), 501–509, 502.

perature *can be explained* in terms of the movements of the molecules, and the temperature of the gas *cannot have a causal force* which is not reducible to the causal forces of the molecules. However, the temperature is clearly an emergent property in Bunge's sense if the 'components of the system' are taken to be the molecules of which the gas consists. This is because the individual molecules do not have a temperature, though their totality does.

With the possible exception of the definition which refers to causal force, each of these definitions can be applied only in a situation in which a detailed description of the relevant 'parts' of the considered system is available. For each definition refers to the properties of such parts and contrasts an imagined detailed description of these properties with a 'macroscopic' description of the same system. In particular, Enqvist's idea that the coarse-grained description was *deducible* from the underlying reality makes sense only when there is *one particular correct description* of the underlying reality.

The difficulties in interpreting quantum mechanics make the assumption that there is such a true description problematic when the theory which is supposed to provide us with the true description is based on quantum mechanics, as e.g. the supposed Theory of Everything would be, and as all contemporary cosmological theories are. The *measurement problem* of quantum mechanics makes it unclear what, precisely speaking, the content of such a detailed description should be.

I shall below give a compact and comprehensible presentation of the measurement problem and of the difficulty that it poses *both* for the various forms of emergentism *and* their denials. The material below contains little that would not be familiar to the physicists and philosophers who are specializing in the foundational problems of modern physics. Yet it seems to me that the point which I make below is worth making, since it is not always noticed by other philosophers. In handling the problems of emergence, philosophers often simply assume that a detailed microphysical description of systems with emergent properties exists without giving any arguments for this assumption.[8] In addition, physicists sometimes fail to give the measurement problem of quantum mechanics sufficient weight in presentations which are aimed at the general public.

In order to clarify the difficulties that the measurement problem causes, I shall contrast the new situation to which quantum mechanics has led with two cases in which the relation between a coarse-grained description and a detailed description of the same reality is quite unproblematic. The latter of these is the example of the temperature of a gas, which was already mentioned above, and also the former is used in Enqvist (*in this volume*) as an analogue of the relation of weak emergence.

---

8 See e.g. A. Beckermann, 'Supervenience, Emergence, and Reduction,' in: Beckermann, Flohr, and Kim (eds.), Emergence or Reduction?, 94–118, and the other essays in the same volume.

2. Two unproblematic examples of weak emergence

In one of Enqvist's examples a time-lapse photograph is taken of a city street. In such a photograph the pedestrians appear as a continuous fog on the pavement, and cars appear as streaks of light, an effect of their headlights.

The coarse-grained description is here a description of the fog and of the streaks of light. What would a detailed description of the same phenomena be like? If we restrict our attention to what is happening on the pavement, we can produce such a description by dividing the pavement into small squares and the period of time during which the time lapse photograph is taken into short intervals. For example, we might choose to divide a part of the pavement into squares which form the rows A, B, C, and D and the columns 1, 2, 3,..., 10 (cf. Figure 1). If the time lapse photograph has been taken during (let's say) the time period from 12am to 12.10am, we might choose to consider time intervals of one second, starting from the interval from 12:00:00 to 12:00:01 and ending with the interval from 12:09:59 to 12:10:00.

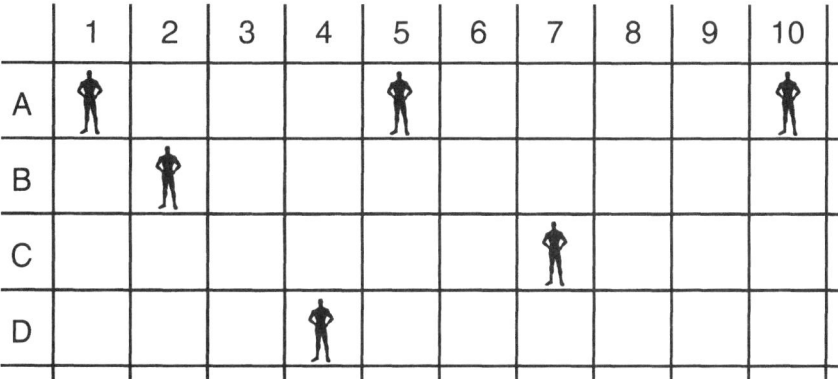

Figure 1.

With this choice of time intervals and the coordinate system for the pavement, a detailed description of reality can be taken to consist of the specifications of the positions that the pedestrians have within the coordinate system at each second. Thus, the detailed description will tell us which squares are occupied by any pedestrian at each second. For example, if the pedestrians are during the second from 12:00:00 to 12:00:01 located as shown in Figure 1, a detailed description will state that during that second of time pedestrians were located at positions A1, A5, A10, B2, C7 and D4. The detailed description will contain altogether 600 different statements of this type, each corresponding to a different second between 12:00:00 and 12:10:00.

In our current example the time-lapse photograph provides a *coarse-grained* description for the phenomenon with which such statements are concerned. If we divide the part of the pavement which is shown in the photograph into squares which correspond to the squares of Figure 1, some of these squares will turn out to look more foggy than others. If we introduce a classificatory scheme for these squares in terms of their 'fogginess', we can get a clear idea in which sense such 'fogginess' is an emergent property. For example, we can divide the squares into 1) the squares which *look very foggy,* 2) the squares which *look relatively foggy,* and 3) the squares which *look foggy only slightly, or not at all foggy.* Now, a coarse-grained description of the phenomenon with which our 600 low-level statements deal can be taken to be one which specifies to which extent each square look foggy. For example, such a description might state that the square A1 looks quite foggy, that the square B1 looks somewhat foggy, that the square C1 does not look foggy at all, and so on.

If we make the simplifying assumption that the 'fogginess' of a square on the photograph depends only on the number of seconds it has been occupied by a pedestrian (which means that, among other things, we are leaving out of consideration the fact that not all pedestrians are of the same size), it turns out that the detailed description and the coarse-grained description are connected by *correspondence rules.* These are rules which specify for each possible coarse-grained description which detailed descriptions correspond to it as, e.g., the following rule does in the case of the 'quite foggy' squares of the photo:

(1) An square x looks quite foggy if and only if it has been occupied by a pedestrian during at least 20% of the seconds between 12:00:00 and 12:10:00.

If similar correspondence rules exist also in the case of the other two alternatives which are allowed by our classification scheme, the macroscopic description is *uniquely determined* by the microscopic one. Together with the laws of optics and the other relevant physical laws, the microscopic description also suffices to *explain* why the macroscopic description is the one that it is, i.e., why the photograph looks the way it does. The fogginess which can be observed in the photograph can be called an emergent property only in the weak sense that it is determined collectively by the behavior of all the pedestrians and in that, accordingly, it cannot be attributed to the individual pedestrians but only to their totality.

It is instructive to contrast the relation between the two descriptions that we are currently considering with the relation which emergent materialists claim to exist between a microphysical description of brain states and entities like beliefs, which are mentioned in the folk psychological descriptions of mental phenomena. An emergent materialist might accept the idea that the mental state of a person, including the beliefs that she has, is uniquely determined by the microphysical state of her brain, analogously to the way in which the fogginess of a square is determined by the number of the pedestrians who have walked on it. Emergent materialists would typically claim, however, that the fact that the person has beliefs cannot be explained by the state of her brain; however hard we study her brain on a microphysical level, we could never according to emergent materialism arrive at an explanation of the fact that she has beliefs.[9]

However, the example of a time-lapse photograph is analogous to another example which we mentioned before, i.e., the example of the temperature of a gas. Within *statistical mechanics* one regularly views an *ideal gas* as consisting of molecules which move along well-defined trajectories and which have well-defined kinetic energies. If one makes such idealizing assumptions, one can show that the average kinetic energy $(E_k)_{ave}$ of the molecules of a gas and its temperature T are proportional to each other. This proportionality provides us with a rule which is similar to correspondence rule (1). When the constant of proportionality is denoted by C, so that $(E_k)_{ave} = CT$, the correspondence rule which according to statistical mechanics applies to an ideal gas is the following rule (2):[10]

(2) The temperature of the gas is T if and only if the average kinetic energy of its molecules is $(E_k)_{ave}$, where $(E_k)_{ave} = CT$.

This rule is quite analogous to rule (1), and it leads to a similar relation of emergence between the detailed and the coarse-grained description of the phenomenon under consideration. The value of the temperature is not an emergent property in the strong sense in which its value would not be explicable with the help of physical laws. It is, however, an emergent property in the weaker sense in which the kinetic energies *determine* it *collectively* and in which it is not a property of any particular molecule of the gas, but only of their totality.

Hence, it easy to make sense of the notion of emergence in the context of classical kinetic theory of gases. Next we shall see why this notion of much more problematic within quantum mechanics.

3. *The measurement problem*

Quantum mechanics is a *probabilistic* theory: it provides probability distributions for the values of various observable quantities but not predictions which uniquely specify what their values will be. In it the state of a physical system is represented

---

9 See e.g. I. Niiniluoto, *Critical Scientific Realism* (Oxford 1999), 21–22.
10 As a matter of fact, the constant C has according to the kinetic theory of gases the value (3/2)k, where k is a physical constant called Boltzmann's constant.

by a mathematical entity which is called the *wave function* of the system and which is often denoted by R. The probability distributions of the observable quantities can be calculated on the basis of the wave function R. For example, if the system under consideration consists of a single elementary particle, its wave function yields (among other things) probability distributions for the *position* of the particle and for its *momentum*. These probability distributions might be concentrated around particular values. E.g., the wave function might yield the result that if the position of the elementary particle is measured, there is a probability 1 that it is found in a specified small region R of physical space, and that the probability that it is found elsewhere is zero. In this case one can say that the wave function represents a particle which is located within the region R.

The measurement problem of quantum mechanics is associated with the enigmatic relation between the wave function R and observable reality. The wave functions of particles might be distributed across large regions of space. Yet when the position of an elementary particle is measured, the particle observedly always has some particular, well-defined location. Such positions can be directly observed, e.g., when a charged elementary particle traverses a bubble chamber, leaving behind itself a sequence of bubbles. Such bubbles are regularly interpreted as showing locations at which the particle has been and has interacted with matter. Physicists suppose that when the particle causes the bubble, it has been localized in the region in which the interaction which causes the bubble has occurred. This region is much smaller than bubble which can be observed in the photo.

We can view the statements which specify the macroscopic, observable locations of a particle as an analogy to the coarse-grained descriptions of our two previous examples, i.e. as an analogy to the fogginess of the various parts of the photograph and to the temperature of an ideal gas. For example, in the situation of Figure 2 it might be the case that the particle causes bubble A at time t_0 and bubble B at time t_1. In this case the description of the observable, macroscopic reality can be taken to contain the statements that at time t_0 the particle was located within region A, and at time t_1 it was located within region B.

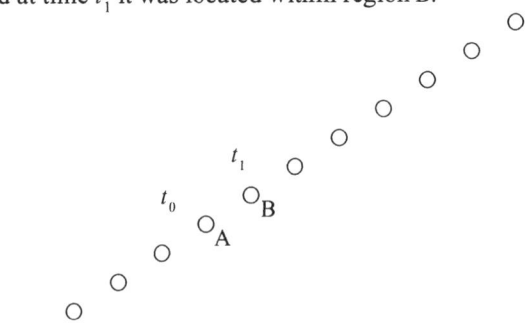

Figure 2.

It is clear that since the description of a physical system in terms of its wave function is mathematically much more complicated than the description which statistical mechanics provides for an ideal gas, the relation between a wave function and a macroscopic description of matter must be mathematically more complicated than the relation which is expressed by the equation in (2). A reader not familiar with the measurement problem of quantum mechanics, however, might nevertheless assume that, in so far the problems of emergence are concerned, the two relations are similar.

One might think that, analogously to formulating the correspondence rule (2), one could formulate rules which state what the wave function R must be like, if a system is to have some particular macroscopic, observable features. If such rules existed, the main difference between them and rule (2) would be philosophically irrelevant: the difference would be simply that the equation which occurs in (2) would in the context of quantum mechanics have to be replaced with a much more complicated mathematical formula. If such rules existed, quantum mechanics and the theories which are based on it would provide the basis for a doctrine that the observable features of matter are emergent in a sense similar to that in which the temperature of an ideal gas is emergent, or in which the fogginess of a time-lapse photograph is emergent.

However, the mathematical formalism of quantum mechanics makes such rules problematic. Problems arise not because there were no wave functions that represent (let's say) a particle which is certainly to be found within bubble A, or within the much smaller region of space A' within bubble A in which the particle has interacted with matter. Rather, the problems arise because of the way in which the wave function according to quantum mechanics changes as a function of time.

Between measurements the wave function of a particle develops according to quantum mechanics in accordance with the deterministic Schrödinger equation. This equation also puts constraints on what the wave function can be at each moment of time. One of these constraints is called Heisenberg's *uncertainty principle.* According to this principle, the precision with which a wave function can yield the value of the position of the particle and the precision with which it can yield the value of its momentum are related to each other: the more precise is the specification of the position, the less precise must be the specification of the momentum, and *vice versa.*[11]

The Schrödinger equation implies also that the region over which the wave function is spread becomes in the long run larger. Hence, if the region within which the wave function is localized at time t_0 is a region A' within the bubble A, the region in which it should be localized according to the Schrödinger equation

---

11 See, e.g., L.I. Schiff, *Quantum Mechanics* (Tokyo 1981), 60–61.

must at some later times be larger than A'. As a matter of fact, the spreading of the wave function is the faster the wider is the range of the momenta with which the wave function was originally compatible.[12] Together with the uncertainty principle, which states that the range of the momenta must be the larger the smaller is the region in which the wave function was originally localized, this fact implies that *the smaller the region in which the wave function was originally localized is, the faster is the process by which it spreads into a larger region.*

The spreading of the wave function is not reflected in the events which trigger bubbles in a bubble chamber: it is not the case that, if bubble B is triggered by an event which happens in region B', this region is typically larger than region A'. Accordingly, physicists do not view the evolution of quantum mechanical systems as consisting merely of a development in accordance with the Schrödinger equation. Rather, they postulate a phenomenon called the *reduction of the wave function*. They assume that during the measurement interaction the wave function of the particle becomes localized in the place where the particle interacts with matter. The location at which this happens cannot be deduced from the wave function which exists before the measurement. Rather, one can calculate on its basis only a probability distribution for this location.

The necessity of postulating a reduction in the wave function makes its status quite problematic. If the reduction of the wave function is viewed as a mere change in our *knowledge* concerning the position of the particle, rather than as a change in the *actual state* of the particle, one is forced to conclude that the quantum mechanical description of the particle is incomplete. In this case the particle has, as a matter of fact, been localized in space also before the measurement, although this location cannot be deduced from the wave function which it at that time had. But if the reduction of the wave function is taken to be an actual physical process, during which something changes in each region of physical space in which the wave function changes, the reduction seems to contradict the theory of relativity. This is because if the particle is observed in one region, according to quantum mechanics its wave function becomes localized in it *immediately,* and disappears immediately everywhere else, although, according to the theory of relativity, it should not be possible that the influence of any event spreads faster than the speed of light.

The claim that the reduction of the wave function is a real physical process is also problematic in another way: it leads to an enigma which in popular literature is often called the problem of *Schrödinger's cat*.[13] The difficulty emerges when one

---

12 Cf. Schiff, *Quantum Mechanics,* 63–64.
13 More precisely, the expression 'Schrödinger's cat' refers to a thought experiment with which the physicist E. Schrödinger illustrated the measurement problem of quantum mechanics. See E. Schrödinger, 'Die gegenwärtige Situation in der Quantenmechanik,' *Die Naturwissenschaften* 23 (1935), 807–812, 823–828, and 844–849, S. 812.

asks *which interactions precisely count as measurements*. We thought above that a measurement happens when an elementary particle causes a bubble to emerge in the bubble chamber. But why could one not view the whole bubble chamber as a single quantum mechanical system which develops in accordance with the Schrödinger equation, and claim that a measurement happens only when someone observes whether bubbles have emerged in the bubble chamber photo? There is nothing in the mathematical formalism of quantum mechanics which rules out this alternative. Yet this choice leads to the seemingly absurd consequence that before anyone observes whether there are bubbles in the photo, there is no truth on the matter whether there are bubbles in it, since the wave function which represents the photo is compatible with both alternatives.

Also more generally, quantum mechanics fails to tell us which interactions should count as the measurements in which the wave function gets reduced. Together with other anomalous features of measurement interactions in quantum mechanics, this failure constitutes the *measurement problem of quantum mechanics*. I shall not try to discuss in such a brief paper the measurement problem in any systematic manner within the context of ordinary quantum mechanics, nor shall I consider the form that the problem would receive within the other theories that are based on the principles of quantum mechanics, such as the Theory of Everything. Rather, I shall relate the measurement problem to my previous discussion of emergence.

The coarse-grained descriptions of our two previous examples contained statements which described the fogginess of various parts of the pavement and statements which described the temperature of the gas. We have already seen what the appropriate analogy of these statements is in the context of Figure 2: the analogy is formed by the statements which specify the positions of the particle at various times. For example, such a statement might claim that the particle was at bubble A at time t_0. Such a statement would describe a weakly emergent property of the system under consideration if it were deducible from a corresponding detailed description of the system. But *what* is the detailed quantum mechanical description of the system, precisely?

If one believes that the reduction of the wave packet is only a change in our knowledge, one is forced to conclude that the correct detailed description of the system under consideration contains elements which its quantum mechanical description fails to mention. In this case one must conclude that the particle was, as a matter of fact, also before the measurement at the place at which it was observed, although its quantum mechanical description did not mention this. On the other hand, if one believes that the reduction of the wave packet is a real physical process, one is confronted with the question in which situations it is precisely that the reduction happens. As the problem of Schrödinger's cat shows, any answer to this question is bound to be somewhat arbitrary. Also somewhat arbitrary, then, are quantum mechanical descriptions of reality. Apparently, unlike in our two previous examples, in the context of quantum mechanics there simply is no one true description of reality, from which one could deduce macroscopic features of reality.

It is perhaps appropriate to emphasize that the source of the trouble is not that quantum mechanics is an indeterministic theory in the sense that in it the knowledge of the state of a physical system at a given moment of time does not suffice to determine its later state. (This indeterminism is, of course, introduced by the reduction of the wave packet; as stated above, between measurements quantum mechanical systems develop in accordance with the deterministic Schrödinger equation.) By itself, the fact that a physical theory is indeterministic does not imply that it cannot yield rules which are similar with our rules (1) and (2).

This point can be illustrated by considering an imagined physical theory which (let's say) resembles the doctrines of the ancient atomists in claiming that elementary particles are solid spheres, rather than the wave functions with which quantum mechanics deals, but resembles quantum mechanics in being indeterministic. The theory might be such that, together with a precise specification of the state of a particle and the state of its environment, it implies that the particle had a number of possible trajectories and specified probabilities for them. In this case the theory would state that the particle will take one of these paths but it would fail to tell us which one it will follow.

It is conceivable that a theory like this could be used for formulating correspondence rules which are similar to (2), i.e. rules which associate macroscopic features of the world with sets of arrangements of elementary particles. Even conceivable is that it could play the role that some emergentist philosophers have thought physical theories would play in the context of the mind-body problem. In other words, it is conceivable that one could specify which states of the human brain, as described by the theory, correspond to the psychological state of having some particular belief. Quantum mechanics makes this idea problematic, but the problem does not arise because of the indeterminism of quantum mechanics but because of the measurement problem. I.e., it arises because of the difficulties that are caused by the reduction of the wave function.

4. Concluding Remarks

We saw above that the definitions of emergence which philosophers have produced are problematic in the context of quantum mechanics, since they do not state what the detailed description of reality which is mentioned in them precisely speaking is in the context of quantum mechanics. But what should one conclude from the existence of this problem?

If one takes the measurement problem seriously, one might conclude that *both* the various forms of emergentism which we earlier mentioned *and* their denials will, when they are applied to the relation between microphysical and macrophysical descriptions of reality, remain merely abstract philosophical positions which the natural sciences can neither support nor refute. This conclusion might be backed up by claiming that the fundamental theories of physics simply do not provide the kind of detailed descriptions of microphysical phenomena which are postulated by both emergentists and their opponents.

This conclusion is, however, in my view based on an unhistorical view of the natural sciences. The development of the natural sciences does not consist only in an accumulation of truths concerning the natural world. Rather, the natural sciences have developed also through radical changes in their fundamental theories. To quote an obvious example, Newtonian mechanics, which was for a long time thought to be the final theory of its field, was replaced in the early 20th century by quantum mechanics and the theory of relativity.

A radical change in fundamental theories can change also the status of problems which have previously seemed insoluble. To take again the obvious example, Newton's theory of gravity failed to answer the question concerning the cause of gravity. Only the general theory of relativity provided a solution to the problem of explaining the cause of gravity. In the general theory of relativity the existence of gravity is explained by appealing to the curved structure of space-time. This solution could not even be formulated within the framework of Newtonian mechanics, however; rather, it became possible to formulate it only after the problems with which the theory of gravity deals had been radically reconceptualized.

Since the problem of explaining gravity was insoluble within the framework of Newtonian physics, physicists before the discovery of the general theory of relativity usually accepted the existence of gravity without trying to explain its cause. Analogously, the lack of progress in solving the problems of interpreting quantum mechanics after its mathematical formalism was discovered in the 1920's seems to indicate that the measurement problem is simply insoluble within the context of contemporary physical theories. The subsequent discussion concerning the measurement problem has produced conceptual clarity concerning the various positions that can be taken towards it, but this discussion cannot be claimed to have produced anything that can reasonably be called a solution to the problem.[14] Accordingly, presently most physicists are unwilling to discuss the measurement problem of quantum mechanics, and view it as a brute fact of nature.

---

14 See J. Bub, *Interpreting the Quantum World* (Cambridge 1997), for a good introduction to the current state of the debate concerning the measurement problem.

Although scientific communities tend to refuse to discuss problems in the context of which their fundamental theories are useless,[15] such problems do not necessarily lack solutions. In particular, the history of science suggests that, instead of viewing the enigmatic features of quantum mechanics as mere facts of nature, one should view them as limitations in the knowledge of physicists at the present stage of science. Also when these limitations are not viewed as mere facts of nature, they have important consequences for both the philosophy of physics and the philosophy of the mind. The enigmatic features of quantum mechanics motivate scepticism within the philosophy of physics towards the early discovery of a final theory of physics, which e.g. the Theory of Everything is supposed to be. In the context of the philosophy of the mind and, more generally, in the context of the philosophical problems of emergence, these features suggest a critical attitude towards all philosophical theories whose validity depends on the details of our current physical theories, and which are likely to become obsolete in the future.

---

15 T.S. Kuhn, *The Structure of Scientific Revolutions* (Chicago ²1970), 37. Here Kuhn points out that science progresses as fast as it does partly because scientific communities are willing to consider only problems that can be solved on the basis of the existing fundamental theories.

10. Time and Causality in the Theory of Everything

Kari Enqvist (Helsinki)

The world around us consists of a bewildering variety of different phenomena. Nevertheless, patterns of weather, the phases of the Moon, the properties of solids, gases, or liquids, reactions between chemical compounds, and the existence of radioactivity, are all but low energy manifestations of a few fundamental interactions. The way these interactions appear to us very much depends on the energies with which we probe the various phenomena, or, equivalently, on the length and time scales we choose to view them. Energy is directly related to resolution: the higher the energy, the smaller the spacelike and timelike distance one can observe. Moreover, as a consequence of the basic laws of physics, in nature there dynamically appear several energy scales related to various structures so that the world can look qualitatively quite different at different resolutions. One could, in fact, go as far as to say that existence itself depends on the energy scale. This, as I will argue, is likely to hold also for space and time.

Solid matter such as wood appears impenetrable only at low enough energies. Visible light cannot pass through doors, but gamma-rays, which are just more energetic light, can do so. Solidity is not a fundamental property of materials but merely the outcome of the various interactions between the molecules as observed with a certain resolution.

Molecules exist due to the residual electromagnetic fields of the nuclei and electrons that leak out of the electrically neutral atoms. As a consequence we get all of chemistry with molecular binding energies – or strengths of molecular bonds – of the order[1] of 0.1 eV; this is the energy per molecule released in chemical burning and roughly speaking, above this energy scale molecules do not exist. For instance, gamma-rays do not 'see' molecules but rather their atomic constituents.

The energy scale associated with atoms is about 10 eV, which is of the order of the binding energies of the atomic electrons. By definition, binding energy is the energy needed to break up a bound state; thus 10 eV is the typical energy needed to ionize atoms. Nuclei, which are bound states of protons and neutrons (collectively called nucleons), have binding energies of a few MeV; this is the

---

[1] I use the units where the Planck constant $h/2\pi$, the speed of light c, and the Boltzmann constant k are all set equal to 1. The basic unit of energy, mass and temperature is then electron volt, eV = 1.6×10^{-19} J, whereas the unit of length is $1/\text{eV} = 2 \times 10^{-7}$ m; time is also expressed in units of $1/\text{eV}$ = 6.6×10^{-16} s. Often used multiples of eV are MeV = 10^6 eV and GeV = 10^9 eV.

energy per nucleon released in a nuclear explosion. Protons and neutrons consist of quarks, which have binding energies of about 1 GeV, which is also the mass scale of nucleons.

The size of an atom is one hundred thousand times the size of the nucleus; the size of the proton is about 1 fm = 10^{-15} m. From various experiments it is known that at these length scales only four fundamental interactions exist: strong interactions, weak interactions, electromagnetic interactions, and gravitational interactions. Strong interactions overcome the electric repulsion between the protons and bind them, together with neutrons, into nuclei. Weak interactions are responsible for certain types of radioactivity, and, for example, the Sun would not shine without them.

Strong and weak forces are very short range forces. Their effects are not felt outside the nuclear length scales. In contrast, electromagnetism has infinite range. Therefore it can act at all length scales. It binds electrons to nuclei as well as atoms into molecules and molecules into compound molecules. Thus, we may thank electromagnetism for the existence of most of the structure important to us, including DNA.

Gravitational interactions also have infinite range and, in a manner of speaking, give us, among other things, the phases of the Moon, but they are very weak and are not important at nuclear length scales. But at the scale of Man and up to the length scales of billions of light years gravity is the dominant player.

Since the 1970's weak and electromagnetic interactions have been known to be in fact two separate low energy manifestations of one single interaction called the electroweak interaction, the energy scale of which is about 100 GeV, which corresponds to the length scale of 10^{-17} m.[2] The theory of electroweak interactions is called the Glashow-Weinberg-Salam model. Its predictions have been verified to a very high accuracy in particle accelerators, although one of its main ingredients, the so-called Higgs particle, is still to be found. Finding the higgs is the main objective of the next large particle accelerator experiment at CERN, the European Center for Particle Physics.

The Glashow-Weinberg-Salam model is an example of a unified theory. It successfully combines two seemingly quite different interactions within a single scheme. Electromagnetism has infinite range whereas weak interactions have a finite, very short range, and at energies of the order of MeV, weak interactions are 100 million times weaker than electromagnetic interactions. Yet at high enough energies they are in fact indistinguishable. Such convergence can happen because in quantum field theories the constants of nature, such as the electric charge of the electron, are not true constants but depend on energy; i.e. on resolution.

---

2   See e.g. S. Weinberg, *Dreams of a Final Theory* (London 1993); G. 't Hooft, *In Search of the Ultimate Building Blocks* (Cambridge 1997).

The Glashow-Weinberg-Salam electroweak theory is a quantum field theory of a specific kind, a so-called gauge field theory. It incorporates all the successes of ordinary quantum mechanics while at the same time providing a generalization of quantum mechanics.

Quarks interact by exchanging quanta of strong force, called the gluons. The theory of strong force is called QCD (Quantum Chromo Dynamics). The nuclear interaction between protons and neutrons comes about because of the small leakage of the gluon field out of the proton or neutron. Like the electroweak theory, QCD is a quantum field theory; it is also a gauge field theory. Because they are theories of the same kind, it could be expected that the strong and electroweak interactions might be low energy manifestations of one single grand unified theory, or GUT. In fact, although at low energies the interaction strengths of the strong and electroweak forces are quite dissimilar, because of quantum effects they actually become equal at a staggeringly high energy scale of 10^{15} GeV, which is thus expected to be the scale of grand unification.

In this picture QCD and the electroweak theory are *effective theories* valid at energies much lower than the GUT energy scale. Their range of validity is restricted, in the same way as the description of matter in terms of molecules is restricted to length scales larger than a nanometer. A molecule is an entity that exists as distinct from a bound state of atoms (and their electromagnetic field, or the collection of electromagnetic quanta – the photons – around them) only when the atoms are seen with a low enough resolution. Molecular bonds – or chemistry in general – are time- and space-averaged representations of the photon exchanges between the constitutents of the atoms. A molecule can be said to be a *coarse grained* description of reality. Chemistry is an effective theory of this coarse grained reality and naturally has a limited range of validity.

Likewise, an atom is a coarse grained description of the dynamics of its constituents, the electrons and the nuclei. Indeed, protons and neutrons are coarse grained descriptions of a multitude of quark-gluon interactions, and so on.[3]

The important point is that coarse graining brings about a qualitatively new description of the same underlying reality: a molecule is a different beast from a mere collection of atoms. This difference comes about because coarse graining results in a description with a lesser number of degrees of freedom. By definition coarse graining takes (weighted) averages of, or integrates out, some of the true dynamical variables of the system.

A time lapse photograph of, say, a busy high street in a city, can serve here as an analogue. People walking on the pavement are blurred into oblivion while cars are present only as streaks of light painted by their headlights. Much of the

---

3   For a popular level account of coarse graining in physics, see e.g. M. Gell-Mann, *The Quark and the Jaguar: Adventures in the Simple and the Complex* (London 1995).

detailed information about the dynamical state of the street scene has been lost in the photograph. Nevertheless, we must concur that it is a valid and a faithful description of that reality, albeit a coarse grained one. It presents us with a new, qualitatively different description of the street scene that instead of individual people has a featureless fog of pedestrians.

Coarse graining leads to *weak emergence*. This means that coarse graining gives rise to descriptions which can be deduced from the underlying reality but cannot be unambiguously reduced back to it. For example, no amount of reasoning based on a time lapse photograph alone can bring back the images of the people into the picture. Generally speaking, this impossibility is a consequence of the fact that an average value of some observable quantity is not a property of any single constituent, although averages can unambiguously be calculated from the constituents. Indeed, in principle we could deduce how a time lapse photograph will look like from the detailed knowledge of the street scene, the shutter speed and the properties of the photographic film. Likewise, the emergent properties of the relatively simple molecules can nowadays be numerically computed starting from the properties of the atoms and from quantum mechanics.

Weak emergence in physics is *dynamical* and not just a feature of our theories (philosophically, emergence encountered in physical systems is not purely *epistemological*). For instance, because of dynamics, the electromagnetic field leaking out of a bound state of several atoms (roughly speaking) lacks wavelengths smaller than the size of the bound state. Well known is that with light one cannot detect features smaller than the wavelength of the light; this fact sets the ultimate limit on the resolution power of an optical microscope. The field that leaks out is the eyes and the fingers of the bound state; therefore it 'sees' the outside world with a certain resolution only. In particular, it does not see its fellow bound states as collections of atoms but rather as coarse grained wholes – as molecules.

Coarse graining as practiced by the molecule is, therefore, a direct consequence of its architecture, which reflects the nature of physical laws.

Temperature is another, perhaps a more familiar example of an emergent quantity in the weak sense of the word. If we look at a large collection of air molecules, none of them can be said individually to possess any temperature. Temperature is the mean kinetic energy of the gas as measured by a thermometer bombarded by these molecules over a period time (thus taking a time average is the act of coarse graining in this case). Temperature does not belong to any single molecule, yet it can be determined from a detailed knowledge of the velocities of the molecules. Several microconfigurations can give rise to the same temperature, so that from a thermometer reading one cannot unambiguously deduce the molecular configuration of the gas.

Because of coarse graining, reality at different length scales appears qualitatively different, as I have discussed. A deterministic theory, when coarse grained, can give rise to an effective theory that is stochastic. Here the familiar example of deterministic chaos in classical physics comes to mind. Alternatively, a stochastic theory, when coarse grained, can give rise to an effective theory that is deterministic. Everything seems to hang on the resolution power of the beholder.

Until recently, our view of space and time has, however, followed a different logic. Matter can be reduced to its constituents; time and space seem continuous. While several coarse grained descriptions of matter are extant, time and space are similar both at very large and at very small scales: no separate time exists for elementary particles and for human beings. I want to argue, however, that in a Theory of Everything coarse graining will necessarily be extended to the space-time itself.

Time and space do not form merely a passive stage upon which elementary particles enter to play their allotted roles. Space and time are dynamical, as described by Einstein's General Theory of Relativity.[4] According to Einstein, gravitational forces are not true forces. Gravity is a coordinate effect: it is how an intrinsically curved space is experienced by an observer who mistakenly believes that we live in a flat space.

General relativity is a field theory but not a quantum field theory. It is a dynamical theory of space and time, which together form a four dimensional manifold, called the 'space-time'. Already in Einstein's Special Theory of Relativity, which describes the kinematical behaviour of particles, space and time are not independent. A well-known example is the running of a clock which depends on the relative velocity of the observer with respect to the clock. Time dilatation is negligible at small relative velocities but becomes noticeable when the velocities are close to the speed of light.

General relativity introduces a new element into the interplay between space and time. Mass and energy affect space-time, which continuously deforms in response to the driving force of mass-energy. Thus space-time is not a static platform but a changing dynamical entity. This action is described mathematically by a field, the space-time metric, which gives rise to such geometrical concepts as the curvature of the space-time. Deep down, the space-time metric field is what, according to general relativity, space and time *really* is.

Apart from the question concerning the nature of time, there is another (although not necessarily a separate) question of why time runs. Why is there an arrow of time? Why can we travel in any direction along the spatial coordinates but only forward along the time-like coordinate? This seems to be the fact despite that most particle physics reactions are time reversal invariant, i.e., if the direction of time is changed, the probability for the reverse reaction is equal to the original. Nature at the level of elementary particles, however, is not completely time

---

4   The standard introduction to general relativity is C. W. Misner, K.S. Thorne and J.A. Wheeler, *Gravitation* (1973 San Fransisco).

symmetric. Theoretically, quantum field theories are invariant under the combined symmetry operations of charge conjugation (C), reflection (P), and time reversal (T), collectively denoted as CPT. CP-symmetry has been long known is violated by the weak interactions, and hence by implication T-symmetry as well. T-symmetry violation has recently been observed directly, too, in certain K-meson decays.[5]

T-violation is very rare, but it is nevertheless present in nature. Unclear is whether this is also the reason for the arrow of time. General relativity does not tell us why time runs, but the reason may be found in particle physics rather than in theories of gravity.

General relativity is deterministic in that equal initial conditions will always lead to the same final state, and it is causal in that no information can propagate locally faster than the speed of light. Under certain circumstances general relativity may allow for closed time loops, but these are unlikely to lead to time travel as we imagine it. Rather than actually travelling anywhere, the observer would just be doomed to repeat everything in exactly the same way all eternity. For even with closed time loops, the solutions to the equations of motion of general relativity should be self-consistent and deterministic for all times. Time travel with wormholes, which are tunnel-like distortions of space and time, seems to require a new, exotic form of matter, which violates the so-called Weak Energy Condition that states that the energy density of matter should be positive.[6] Even if such new type of matter exists, the wormholes are likely to be unstable against quantum fluctuations. Stephen Hawking has elevated these considerations to a chronology protection conjecture, according to which the laws of physics do not allow the appearance of closed timelike curves.[7]

Because of its extreme weakness at microscopic scales, gravity is neglected in the strong and electroweak quantum field theories. Their space-time background is that of the Special Theory of Relativity, and they are causal in the sense that no information propagates faster than the speed of light. The causally connected spatial region is called the light cone. No faster-than-light propagation has ever been observed, except for the recent claims that signals have been observed to quantum mechanically tunnel through energy barriers with several times the velocity of light. The results of these experiments have, however, probably been misinterpreted.[8]

---

5 Apostolakis et al., 'Determination of the T and CPT Violation Parameters in the Neutral Kaon System Using the Bell-Steinberger Relation and Data from CPLEAR,' *Physics Letters* B456, (1999), 297–303.
6 See, e.g., K.S. Thorne, *Black Holes & Time Warps* (New York 1994).
7 S.W. Hawking, 'Chronology Protection Conjecture,' *Physical Review*, (1992), D46, 603–611.
8 For a non-technical discussion of faster-than-light signals in quantum mechanics, see R. Chiao, P.G. Kwiat and A.M. Steinberg, 'Faster than Light?,' *Scientific American*, (August 1993), 38–46; J. Brown 'Faster than the Speed of Light,' *New Scientist*, (1 April 1995), 26–30.

As quantum theories, the strong and electroweak theories are not deterministic: equal initial conditions do not always lead to the same final state. A given final state will be obtained only with some calculable probability.

The probabilistic nature of quantum physics is an inherent feature of the theory. Quantum physics does not give a statistical description of many particles but a probabilistic description of a single particle. Coarse graining can be applied also to quantum mechanical systems, but it cannot remove the probabilistic nature of the theory.

Formally, quantum field theories can be described in terms of correlators, just like classical statistical theories, albeit in the case of quantum theory the correlator is a complex rather than a real valued function. A correlator is a function from which one may read the probability amplitude for the following: given that the value of the field at the space-time point x is A, what is the probability that its value at y is B? (This represents a two-point correlator; a three-point correlator would answer the question: given that the value of the field at the space-time point x is A, what is the probability that its value at y is B and its value at z is C? A general correlator is an n-point correlator.) Causality in quantum field theories means that all correlators vanish by construction outside the light cone.

We have some compelling evidence that the probabilistic nature of quantum physics is not just a reflection of our ignorance. This fact became evident after the celebrated experiment of Alain Aspect and his collaborators in 1982.[9] Using split photon beams, they measured correlations between photon polarizations at different, spatially distinct polarizers. They found that there was more correlation between the photon polarization states than predicted by local classical physics and probability theory. This shows that, no matter what the ultimate theory is, reality itself is not classical (or local). The measurements were, however, in agreement with the predictions of quantum mechanics, so that it appears that quantum physics captures reality with at least a reasonably good approximation.

Because of the huge success of quantum physics in general, many physicists believe that general relativity is just an effective theory valid only at low energies and that gravity should as well ultimately be described in terms of a quantum field theory. This means that the space-time metric itself should somehow become probabilistic. Instead of a continuous space-time metric field, we should speak only about correlators between various events. The correlators would answer questions such as 'given that a clock at x strikes midnight, what is the probability that at y it is noon?' The many attempts to construct a quantum theory of gravity, however, have all failed, because the General Theory of Relativity and quantum physics are not consistent with each other.

---

9   A. Aspect, P. Grangier and C. Roger, 'Experimental Realization of Einstein-Podolsky-Rosen-Bohm Gedankenexperiment: A New Violation of Bell's Inequalities,' *Physical Review Letters* 49 (1982), 91–94.

According to quantum theory, at very small distances particles are not solid objects but rather fuzzy 'probability waves'; the measure of the spatial extension of the particle is the inverse of its energy. According to general relativity, however, the radius of the event horizon of a black hole can actually be smaller than the spatial extension of the quantum black hole. This happens at energies (masses) of the order of the Planck mass, or 1.2×10^{19} GeV. The event horizon is the radius beyond which nothing can escape the black hole; but then how can a quantum black hole contain itself within the event horizon if its fuzziness is spread over distances larger than the event horizon? Clearly either general relativity or quantum physics has to be modified at energy scales larger than the Planck mass, or at length scales smaller than about 10^{-35} m (i.e. the Planck length). In a quantum theory of gravity space-time itself should, in some sense, become 'fuzzy', or not well defined. This fuzziness would reflect the existence of an indivisible quantum of space-time.

The quantum nature of space-time would have been manifest for the universe as a whole at its birth. At the big bang, when time and space began at the initial singularity, matter was in a state of extremely high density. Applying our knowledge from particle physics experiments, we can trace back the evolution of the universe to a time of about 10^{-11} s after the big bang. This corresponds to the energy scale we have reached with large particle accelerators. We believe that we can discuss theoretically even the earlier moments, but as we get closer to the initial singularity, matter becomes denser and denser. As a consequence, the gravitational effects become strong and can not be neglected for times before the so-called Planck time, or 10^{-43} s. Right after the big bang, before the Planck time, we would need quantum gravity to describe our universe. But in a fuzzy quantum space-time, sometimes called space-time foam, time and space would loose their meaning and so would the notion of the universe as we know it. In a quantum version of the big bang theory, time did not begin at an initial singularity. Rather, it emerged from the fluctuating foam that could be said to consist of all possible universes – or none of them.

A theory which would incorporate both quantum physics and general relativity would in effect unify all of the known physical interactions within a single framework; hence the name Theory of Everything, or TOE.[10] At present the best (and the only) candidate for a TOE is string theory, according to which particles are not really pointlike (albeit fuzzy) objects but rather vibrations of open or closed strings. In string theories space-time itself is just one of the vibrational modes of the string, and strings themselves are typically 10-dimensional objects which live on the surface of a two-dimensional world. (Recently it has been ar-

---

10 Like GUT, TOE is an acronym chosen with tongue in cheek. Its first appearance in print is probably the 1986 *Nature* 323, (595) article 'The Superstring – Theory of Everything, or of Nothing?' by the CERN theorist John Ellis.

gued that string theories may be incidences of an even more abstract theory called the M-theory). The only parameter of a string theory is the string tension α, which is expected to be of the order of the Planck mass.[11]

Under normal circumstances, strings are very small with sizes of the order of the Planck length. There are only five mathematically consistent string theories. They predict gravity in the sense that they automatically accommodate among their vibrational modes a massless particle whose couplings at long distances are those of the corresponding metric space-time field of general relativity. At distance scales much larger than the Planck length the effective theory of strings is a certain type of a quantum field theory plus general relativity. This effective theory describes pointlike particles (fields) which propagate through a curved 10-dimensional space-time. Six of the dimensions are expected to be curled up so that they form something like a generalized 6-dimensional ball with a radius of the order of the Planck length. The resolution required to 'see' the extra dimensions could only be achieved with energies close to the Planck mass scale. Thus at low enough energies, or large distances, it is possible to recover our usual four dimensional world and its pointlike particles as coarse grained descriptions of the true dynamical variables, the strings.

The string nature is however revealed at small distances. Even in the absence of quantum fuzziness, space-time looses its meaning in string theories at length scales of the order of Planck length and at time scales of the order of the Planck time. At scales smaller than these ordinary space-time is no longer 'visible'.

Heisenberg's uncertainty principle in quantum physics states that the spatial resolution Δx and the momentum resolution Δp cannot both be made arbitrarily small at the same time; rather $\Delta x > h/\Delta p$. At very high energies, the energy of the particle is practically equal to its momentum. Hence, as stated, improving spatial resolution requires more energy. It has however been argued that there exists a minimum size for the string so that by increasing energy beyond the Planck mass one no longer can improve on the spatial resolution. In effect, roughly speaking, Heisenberg's uncertainty principle gets replaced by an expression of the type

$$\Delta x > h/\Delta p + \alpha \Delta p/h,$$

where the first term is the usual quantum uncertainty, and the second is the new string induced fuzziness.[12] At low energies the first term on the right hand side

---

[11] For recent popular books discussing string theory, see e.g. B. Greene, *The Elegant Universe: Superstrings, hidden dimensions, and the Quest for the Ultimate Theory* (New York 1999); M. Kaku, *Introduction to Superstrings and M-Theory* (New York 1999). For a more technical discussion, see M. B. Green, J.H. Schwarz and E. Witten, *Superstring Theory* Vols. I–II. (Cambridge 1987); J. Polchinski, *String Theory* Vols. I–II (Cambridge 1998).

[12] E. Witten, 'Reflections on the Fate of Spacetime,' *Physics Today* (April 1996), 24–30.

dominates so that by increasing energy, or momentum Δp, one will increase the resolution Δx. However, at high enough an energy scale, at the string scale, the second term overcomes the first, and the resolution starts to decrease with energy. String theories have in this sense a minimum length, the scale of which is given roughly by the Planck scale, although a mathematical proof is still lacking.

We do not yet know whether string theory, or M-theory, is really the much hoped-for TOE. Nevertheless, it is evident that if quantum physics and general relativity will eventually be brought together, space-time itself should emerge as a low-energy coarse grained approximation of the more fundamental, underlying reality. In the Theory of Everything neither space nor time is likely to be any more fundamental than, say, a molecule. They are just effective, emergent qualities of reality and manifest themselves at low enough energies. Time may thus not exist at the most fundamental level, nor does time ordering. Causality as we know it – a time ordered list of events – would, therefore, suffer a similar fate. Instead of space and time, instead of before and after, events would take place in some highly abstract reality. Nevertheless, there would still be some order in these abstract microevents which at large enough distances would look like causal chains.

What, then, is the true nature of time? At present we cannot tell, but whatever it turns out to be, it will most likely be nothing that would be familiar to us from our everyday experience. The world we observe through our senses is a coarse grained approximation of reality. Because of the coarse graining, it seems to consist of observables that are qualitatively different from the observables at the most fundamental level, i.e. at the smallest distance scales. Therefore, no amount of philosophical reasoning is unlikely to make sense of time.

It is thus possible that time remains a concept which is intimately familiar to us and yet fundamentally alien. If TOE is ever found, time may eventually be dissected and explained mathematically, but it is not inconceivable that there will be nothing for us to understand.

11. The Concept of Space in Relativity Theory and Cosmology[1]

Raimo Lehti (Helsinki)

1. Before Einstein

1.1. The Principle of Relativity in Classical Physics

Galileo, in his *Dialogue Concerning the Two Chief World Systems* of 1632, had already dealt with the relativity of motion. His task was to advance arguments for the 'nonoperativeness' of Earth's rotation: its possibility, he argues, would not produce catastrophic effects upon the Earth that had traditionally been thought. According to Galileo, such nonoperativeness was characteristic of any uniform motion common to the observer and his environment: it does not produce any effect which can distinguish it from the state of rest. Galileo's 'principle of relativity' was a 'conspiracy theory' in so far as it accepted the existence of absolute motion but argued that the laws of nature had conspired to make such motion unobservable. Experiments from the perspective of Earth could never disclose it, since the phenomena are identical for a system of rest and a system of uniform motion. Isaac Newton reacted to the same problem in his discussion of the possibility of defining 'motion' as a change of absolute place when we can observe only relative positions. The Scholium to the Definitions in Book I of Newton's *Principia* includes a well-known and oft commented text where he affirms the existence of absolute space and time. Newton, like Galileo, accepted the 'conspiracy theory of natural laws'. We formulate Newton's principle of relativity as follows:

When studying motion as the function of time t, we specify the location of a body using orthogonal coordinates x, y, z. We may also use another system of coordinates x', y', z', to specify the location of a body moving relative to the first system with a constant velocity v. For simplicity, we assume that the axes of the systems are parallel to each other and that the relative motion takes place in the x - direction. We then have:

(1) $\quad x' = x - vt, \quad\quad y' = y, \quad\quad z' = z, \quad\quad t' = t.$

---

[1] A Finnish-language version of this paper appeared in the number 5/2000 of the journal *Tieteessä tapahtuu*, published by Federation of Finnish Learned Societies.

The assumed existence of absolute space implies that both systems cannot be 'true' in the sense that they would give the 'real' position of a body. Time t is identical in both systems, so this fourth coordinate cannot be used to draw any conclusion about which spatial coordinates should be taken as 'true' and which are moving relative to the true system. According to the classical principle of relativity, the laws of mechanics and the phenomena deduced from them are identical whichever coordinate system is used.

The study of optical and electromagnetic phenomena gave rise to further questions concerning the relativity of motion. The classical theory developed by James Clerk Maxwell and others assumes that the 'carrier' of electromagnetic fields is a universal medium called *ether*. This theory rests upon the assumption that in a coordinate system stationary relative to the ether the electric and magnetic fields satisfy a system of equations called *Maxwell's equations*. This assumption seems to imply that using electromagnetic phenomena might make it possible to decide which coordinate system is 'correct' in the sense that in that system the phenomena occur according to Maxwell's equations. We might by using that coordinate system at last observationally specify the absolute space which had remained a dilemma since Newton. The investigations concentrated especially upon the problem concerning whether Earth's velocity w might influence such phenomena. To decide the question, several experiments were made; but they invariably gave a 'zero result': no difference in the phenomena evoked by Earth's motion was detected. A weakness of the experiments was due to the fact that Earth's velocity w is minute compared to the velocity c of light, which was identified as electromagnetic wave motion. The precision of the experiments allowed in general only the conclusion that no 'first order effects' were found, i.e., effects proportional to the ratio w/c.

Later special weight was given to the attempt to measure the assumed change in the observed velocity of light due to the annual motion of the Earth. The experiment was first done by Albert Abraham Michelson in 1881, and again later with William Morley in 1887. The precision of the latter experiment was sufficient to reveal even a 'second order effect', i.e., an effect proportional to the square $(w/c)^2$. No effect was observed; the experiment again gave a zero result. An investigator of Maxwell's theory, George Francis Fitzgerald, explained the result in 1889:

> that the length of material bodies changes, according as they are moving through the ether or across it, by an amount depending on the square of the ratio of their velocities to that of light. We know that electric forces are affected by the motion of the electrified bodies relative to the ether, and it seems a not improbable supposition that the molecular forces are affected by the motion, and that the size of a body alters consequently.[2]

---

2   Abraham Pais, *'Subtle is the Lord...': The Science and the Life of Albert Einstein* (Oxford 1983), 122.

1.2. Hendrik Antoon Lorentz

The Dutch physicist Hendrik Antoon Lorentz wrote in an 1892 article that he had for a time been deeply worried about the Michelson-Morley experiment. Lorentz was unaware of Fitzgerald's explanation, but he arrived independently at the same conclusion. He also could find only one solution for the zero result: namely that the measuring instrument contracts when moving in the direction of the motion. To explain this contraction, Lorentz assumed that the forces between the molecules of the measuring stick spread through the ether in the same way as the electromagnetic forces; thus the motion causes a change in the molecular distances. Again in 1895 Lorentz discussed the possibility of uniting the result of the Michelson-Morley experiment with other results concerning the velocity of light. Lorentz had elaborated a theory to explain the structure of matter known as *the theory of electrons*. The credibility of this theory required an explanation for the failure of earth's motion to have any effects.[3] Lorentz suggests that, because of Earth's motion, the arm of the interferometer parallel to the motion becomes shorter than the arm orthogonal to that motion. Lorentz writes:

> Surprising as this hypothesis may appear at first sight, yet we shall have to admit that it is by no means far-fetched, as soon as we assume that molecular forces are also transmitted through the ether, like the electric and magnetic forces, of which we are able at the present time to make this assertion definitely...[4]

In 1899 Lorentz published a new exposition concerning the same problem: *Théorie simplifiée des phénomènes électriques et optiques dans des corps en mouvement*. This time he handled the dynamic contraction hypothesis as if it were a coordinate transformation affecting the x-coordinate parallel to the motion. Lorentz combined this transformation with another device which he had used when mathematically manipulating Maxwell's equations. He had recognized that these equations can be conveniently handled in a moving system of coordinates if one introduces instead of the time variable t another variable t', which is a function of the spatial position of the point in question and of the velocity of the moving system. The transformation equations he gave for time and location were, excluding an arbitrary factor, equivalent with those he later presented in his better-

---

3   Russell Mc'Cormmach, 'Lorentz, Hendrik Antoon,' in: Charles Coulston Gillispie (ed.), *Dictionary of Scientific Biography 8* (New York 1981), 487–500, 495.
4   Hendrik Antoon Lorentz, 'Michelson's Interference Experiment,' in: Lorentz, Einstein, Minkowski & Weyl with notes by Sommerfeld, *The Principle of Relativity: A Collection of Original Memoirs on the Special and General Theory of Relativity* (New York 1952), 1–7, 6.

known text of 1904 and which are today known as the equations of the *Lorentz-transformation*.⁵ In their final form the equations are:

(2) $\quad x' = (x - vt)/(1 - v^2/c^2)^{1/2}, \quad t' = (t - vx/c^2)/(1 - v^2/c^2)^{1/2}, \quad y' = y, \quad z' = z$

Lorentz also stated that the contraction is a consequence of these transformations.⁶ The most essential difference between the transformation equations (1) and (2) is that in equations (2) the time coordinate is also changed. Lorentz himself interpreted the 'time' t' only as an auxiliary mathematical device, but Einstein later gave another interpretation.⁷

In his magnum opus of 1909, *The Theory of Electrons*, Lorentz arrives at transformation equations essentially equivalent with (2), creating a 'two-step transformation' consisting of a preliminary 'classical' transformation (1) and a second transformation leading to (2).⁸ The new coordinates x', y', z' Lorentz calls *effective;* by their means he defines the effective distance, effective volume element dS', effective density of the charge ρ', etc.⁹ Denoting the stationary and the moving system with the symbols S_0 and S, Lorentz writes:

> ...That which we have now made enables us to condense into few words what was said in the last paragraph about the systems S_0 and S, namely: In two electrostatic systems, the one moving and the other at rest, in which the effective density of the electric charge is the same function of the effective coordinates, the vector [of the electric field] will be the same at corresponding points, and the forces will be related to each other...¹⁰

Lorentz begins now to use the attribute *effective* for the quantities of both the stationary and the moving system and remarks that in using this terminology the electrostatics of the two systems display no difference. This may lead to the question whether it is reasonable to call some effective quantities the *true* ones. Lorentz, however, believes one may do so; hence, his theory is a new specimen of

---

5   Mc'Cormmach, 'Lorentz, Hendrik Antoon,' 496.
6   Pais, *'Subtle is the Lord...',* 126.
7   I discuss this question more extensively (in Finnish) in my 'Aika suhteellisuusteoriassa ja kosmologiassa' (Time in Relativity Theory and Cosmology), in: Pihlström, Siitonen, Vilkko (eds.), *Aika* (Helsinki 2000).
8   Hendrik Antoon Lorentz, *The Theory of Electrons* (Dover 1952), 196–197, and Hendrik Antoon Lorentz, 'Electromagnetic Phenomena in a System Moving with any Velocity Less than that of Light,' in: Lorentz, Einstein, Minkowski & Weyl with notes by Sommerfeld, *The Principle of Relativity: A Collection of Original Memoirs on the Special and General Theory of Relativity* (New York 1952), 9–34, 14.
9   Lorentz, *The Theory of Electrons*, 200–201.
10 Lorentz, *The Theory of Electrons*, 201.

a 'conspiracy theory': there is a true system, but we are unable to disclose it. Comparing the 'true' and the 'effective' coordinates and supposing that the Lorentz-invariance holds for all molecular forces in the same way as for the electrostatic forces, Lorentz again obtains the result that the dimensions of a body change in the way the Michelson-Morley experiment requires.[11] Accordingly, he writes:

> It will be clear by what has been said that the impressions received by the two observers A_0 and A would be alike in all respects. It would be impossible to decide which of them moves or stands still with respect to the ether, and there would be no reason for preferring the times and lengths, measured by the one to those determined by the other, nor for saying that either of them is in possession of the 'true' times or the 'true' lengths.[12]

Like many other writers on the problems of relativity, also Lorentz identifies here the statement that *certain equations of the general theory and their implications (and, hence, the general laws of nature)* are the same for all observers, with the statement that the impressions the observers get from the world are the same. We will repeatedly meet this question. Although the general laws would place in equal position all observers moving uniformly relative to each other, that situation *does not necessarily imply* that the concretization of the general laws is such that in our real and unique existing world equality would prevail among the observers. This follows simply from the fact that the general laws do not alone determine the structure of the world and that the initial values are also of importance. This fact becomes crucial when we discuss the concept of space in the context of cosmology.

2. Einstein's Special Theory of Relativity

2.1. The Fundamental Problem of Space

Taken generally for granted before Einstein was that the laws of mechanics remain invariant in the transformation (1), which Philipp Frank in 1909 christened the *Galileo-transformation*.[13] No trace of evidence against the Galileo-invariance existed in mechanics in 1905. A conflict arose when the Galileo-invariance was elevated to a universal principle, valid also in electrodynamics and optics. This was the central problem for Albert Einstein's fundamental paper of 1905: *Zur Ele-*

---

11 Lorentz, *The Theory of Electrons*, 202.
12 Lorentz, *The Theory of Electrons*, 229.
13 Pais, *'Subtle is the Lord...'*, 140.

ktrodynamik Bewegter Körper.[14] Einstein takes up the problem in the introduction to the article that experiments with electromagnetic phenomena had not disclosed any effect due to Earth's motion. They rather indicate that the same laws of electrodynamics and optics hold in all inertial systems of reference. Einstein continues:

> We will raise this conjecture (the purport of which will hereafter be called the 'Principle of Relativity') to the status of a postulate, and also introduce another postulate which is only apparently irreconcilable with the former, namely, that light is always propagated in empty space with a definite velocity c which is independent of the state of motion of the emitting body. These two postulates suffice for the attainment of a simple and consistent theory of electrodynamics of moving bodies based on Maxwell's theory for stationary bodies.[15]

Einstein published in 1916 the well-known exposition *Über die spezielle und die allgemeine Relativitätstheorie (Gemeinverständlich)*. There, in paragraph 3, entitled *Space and Time in Classical Mechanics*[16], Einstein accepts with some reservation as the task of mechanics the investigation of how bodies change their location in space. He explains his reservations as follows;

> It is not clear how we are here to understand 'place' (*Ort*) and 'space' (*Raum*). I am standing at the window of a uniformly advancing railway wagon and let a stone fall down the railroad bank, without giving it any impulse. Then I see how the stone falls directly downward (ignoring the effect of air resistance). A pedestrian looks at my offence from the sidewalk and notes that the stone falls to the ground on a parabolic curve. Now I ask: Do the 'places' through which the stone falls 'in reality' lay on a straight line or on a parabola? Moreover, what does the motion 'in space' signify here? ... For the time being we pay no attention to the vague word 'space'; being quite honest, we cannot attain it any real meaning. We replace it with 'motion relative to some practically rigid body of reference'. ... When we next introduce instead of the 'body of reference' the concept of a 'coordinate system', convenient for mathematical description, we may say: The stone describes a straight line in reference to a coordinate system rigidly connected with the wagon; but it describes a parabola, when referred to a coordinate system rigidly connected with the

---

14 Albert Einstein, 'On the Electrodynamics of Moving Bodies,' in: Lorentz, Einstein, Minkowski & Weyl with notes by Sommerfeld, *The Principle of Relativity: A Collection of Original Memoirs on the Special and General Theory of Relativity* (New York 1952), 35–65.
15 Albert Einstein, 'On the Electrodynamics of Moving Bodies,' 38
16 Albert Einstein, *Über die spezielle und die allgemeine Relativitätstheorie (Gemeinverständlich)* (Braunschweig 1920), 6–7.

ground. From this example we clearly see that there does not exist an orbit as such but only an orbit relative to some specified body of reference.

From the logical point of view Einstein makes a somewhat too strong assertion. His example does not prove the non-existence of 'absolute space', neither the lack of meaning of 'motion in space'. Both Galileo and Newton would have understood and accepted the example given by Einstein, but in spite of that they accepted also the concepts of absolute space and motion.

2.2. *The Principle of Relativity*

Einstein formulates in his article of 1905 as formal postulates the two principles he gave already in the introduction.[17] From these principles, the principle of relativity and the principle of the constant velocity of light, Einstein deduces the coordinate transformation (2), which was later called the *Lorentz transformation.* Einstein's method of deducing the transformation equations differs from that of Lorentz; Einstein does not refer to Maxwell's equations or their invariance. He begins with an analysis of *simultaneity* and emphasizes the point that in classical physics such an analysis has been neglected. His procedure gives to the new theory an *epistemological* flavour which many of his followers have taken as an essential feature of the theory, especially one which distinguishes it from the theory of Lorentz. In the interpretation of Lorentz, the parameter t' of the equations (2) was only an auxiliary device; but, according to Einstein's analysis of simultaneity, it was physically fully equal to the 'original' time t. This furnished the *relativity theory* as formulated by Einstein, the dependence of *time* on the system's state of motion, the feature that most disturbed his contemporaries. In this paper I concentrate on *space*[18], although the equations (2) show that space and time cannot be separately treated in a physics which accepts those equations as fundamental.

When equations (2) are compared with the earlier Galileo transformation equations (1), we realize especially one essential difference. Using the earlier transformations, all uniformly moving systems were consistent with the same *time* coordinate t. In equations (1) time remains 'absolute'; but this does not give any clue which of the coordinates x, x' were 'more true'. Therefore, space can be 'non-absolute' even when time is absolute. In contrast, if in using equations (2) we *for some reason* find time t 'more true' than time t', then such a position would lead automatically to accepting as well the spatial coordinate x as 'more true' than x'. *Fixing the time-coordinate* in the Lorentz-transformation (2) leads necessarily

---

17 Einstein, 'On the Electrodynamics of Moving Bodies,' 41.
18 For time, see Lehti, 'Aika suhteellisuusteoriassa ja kosmologiassa.'

also to *fixing the spatial coordinates* (a purely spatial coordinate transformation would, of course, be allowed). This fact implies that if we in some sense accept equations (2) as more correct than equations (1), *then the supposition of absolute time leads to the supposition of absolute space*. Either all 'times' and all 'spaces' are equivalent or there exists a preferred time and a preferred space. We will return to these implications in the cosmological context.

Einstein writes in the exposition of 1916:

> If the principle of relativity (in restricted sense) does not hold, then the Galilean coordinate systems K, K', K'', etc., moving uniformly relative to each other, are not *equivalent* for the description of natural phenomena. Hence, we could scarcely avoid the conclusion that it must be possible to formulate the laws of nature in a specially simple and natural way only if we choose as the system of reference among all Galilean coordinate systems one (K_0) having a definite state of motion. It would be justified to call this one the 'absolutely stationary one' (because of its advantages for describing the nature) and the other Galilean systems as moving.[19]

Then Einstein refers to the zero results of the attempts to discover the effects of Earth's motion and draws upon this argument for the validity of the principle of relativity. We meet here again the transition common to writers about relativity. Einstein argues that if *natural phenomena* with their specially simple appearances suggest a special position of some coordinate system, then the *laws of nature* should also take a simpler form in that system. Einstein defines *absolute space* as that system where the *phenomena* are simple; therefore, the laws should also be simple. The principle of relativity guarantees then, that such a system does not exist. Such a principle is, however, in no way necessary for the special theory of relativity *as a mathematical theory,* although it seems to be essential for most philosophical interpretations of the theory.[20] Even if we accept the 'relativity principle', it makes a difference whether it is formulated in terms of *natural events* or in terms of *natural laws*. The events have other determinants than laws, namely the initial values. Whether or not we accept the principle of relativity for laws, modern cosmology gives a strong argument that it does not hold for natural phenomena.

Einstein argued in 1905 that the ether and absolute space are unnecessary and implausible concepts to be used in a consistent electrodynamics. All his life Lorentz saw the absolute time and absolute space as meaningful concepts.[21] He wrote in 1913:

---

19 Einstein, *Über die spezielle und die allgemeine Relativitätstheorie,* 9–10.
20 See also Albert Einstein, *Out of my Later Years* (Secaucus, N.J. 1956), 277.
21 Mc'Cormmach, 'Lorentz, Hendrik Antoon,' 498.

According to Einstein, it has no meaning to speak of motion relative to the aether. He likewise denies the existence of absolute simultaneity.

It is certainly remarkable that these relativity concepts, also those concerning time, have found such a rapid acceptance.

The acceptance of these concepts belongs mainly to epistemology. ... As far as this lecturer is concerned, he finds a certain satisfaction in the older interpretations, according to which the aether possesses at least some substantiality, space and time can be sharply separated, and simultaneity without further specification can be spoken of. ...[22]

Lorentz can be taken to support the 'conspiracy theory of natural laws', according to which the laws do not make any difference between moving and stationary reference systems, although the difference exists *as a factual feature of our world.* Many have judged the attitude of Lorentz as a sign for that he did not understand the theory of relativity. Such an accusation confuses intellectual comprehension with emotional acceptance. Is it not sufficient to understand a theory as a mathematical tool and accordingly be able to use it? Must it also be accepted with a firm conviction as 'true'? How are we to react to the fact that, as will soon be explained, space and time are again 'sharply separated' in the present cosmology? Must we, perchance, pace the philosophers of relativity, return back to Lorentz's position?

2.3. *Minkowski's Theory of Space and Time*

The mathematician Hermann Minkowski presented the most radical rejection of Lorentz's position in a lecture about space and time, held in Cologne in 1908.[23] The auditorium was startled by the lecture's beginning and end. It begins as follows:

The views of space and time which I wish to lay before you have sprung from the soil of experimental physics, and therein lies their strength. They are radical. Henceforth space by itself, and time by itself, are doomed to fade away into mere shadows, and only a kind of union of the two will preserve an independent reality.[24]

---

22 Pais *'Subtle is the Lord...'*, 166–167.
23 English translation: Hermann Minkowski, 'Space and Time,' in: Lorentz, Einstein, Minkowski & Weyl with notes by Sommerfeld: *The Principle of Relativity: A Collection of Original Memoirs on the Special and General Theory of Relativity* (New York 1952) 73–91.
24 Minkowski, 'Space and Time,' 75.

Minkowski begins the text with an exposition of the invariance property of Newtonian mechanics, or 'Galileo-invariance', as we say today. From mechanics we cannot conclude whether or not the space taken as stationary is in reality in a uniform translational motion:

> ...Let x, y, z be rectangular co-ordinates for space, and let t denote time. The objects of our perception invariably include places and times in combination. Nobody has ever noticed a place except at a time, or a time except at a place. But I still respect the dogma that both space and time have independent significance. A point of space at a point of time, that is, a system of values x, y, z, t, I will call a *world-point*. The multiplicity of all thinkable x, y, z, t system of values we will christen the *world*. ... We fix our attention on the substantial point which is at the world point x, y, z, t, and imagine that we are able to recognize this substantial point at any other time. Let the variations dx, dy, dz of the space co-ordinates of the substantial point correspond to a time element dt. Then we obtain, as an image, so to speak, of the everlasting career of the substantial point, a curve in the world, a *world line*, the points of which can be referred unequivocally to the parameter t from $-\infty$ to $+\infty$. The whole universe is seen to resolve itself into similar world lines ...
>
> ... we may – also without changing the expressions of the laws of mechanics – replace the x, y, z, t, by $x - \alpha t, y - \beta t, z - \gamma t, t$, with any constant values of α, β, γ. ...[25]

Minkowski next gives somewhat a priori reasons for the replacement of the Galileo-transformation (1) with the Lorentz-transformation (2). He demonstrates the possibility of defining such a geometrical structure in the four-dimensional world, i.e., that every 'time-axis' has in a generalized sense an 'orthogonal' complement which can be taken as 'space' corresponding to that 'time'. The Lorentz-transformations can be interpreted in this geometric picture as some sort of generalized 'rotation' from the coordinate axes of one 'space-time-system' to the axes of another one. Having completed this geometric construction, Minkowski writes:

> For example, in correspondence with the figure described above, we may also designate time t', but then must of necessity, in connection therewith, define space by the manifold of the three parameters[26] x', y', z', in which case physical laws would be expressed in exactly the same way by means of x', y', z', t' as by means of x, y, z, t. We should then have in the world no longer *space* but

---

25 Minkowski, 'Space and Time,' 76–77.
26 Minkowski, 'Space and Time,' 79 we have in fact x', y, z, instead of x', y', z'.

an infinite number of spaces analogously as there are in three-dimensional space an infinite number of planes. Three-dimensional geometry becomes a chapter in four-dimensional physics. Now you know why I said at the outset that space and time are to fade away into a shadow, and only a world in itself will subsist.[27]

Minkowski next demonstrates that the hypothesis leading to Lorentz-transformations (2) is completely equivalent to the new conception of space and time, 'which, indeed, makes the hypothesis much more intelligible'.[28] He writes:

> Lorentz called the t' combination of x and t the local time of the electron in uniform motion, and applied a physical construction of this concept, for the better understanding of the hypothesis of contraction. But the credit of first recognizing clearly that the time of one electron is just as good as that of the other, that is to say, that t and t' are to be treated identically, belongs to A. Einstein. Thus time, as a concept unequivocally determined by the phenomena, was first deposed from its high seat. ... Since the postulate [of relativity] comes to mean that only the four-dimensional world in space and time is given by phenomena, but that the projection in space and time may still be undertaken with a certain degree of freedom, I prefer to call it the *postulate of the absolute world* (or briefly, the world-postulate).[29]

We recognize in Minkowski's texts the same conceptual slippage already mentioned. All what he indeed demonstrates is that time has been 'deposed' from certain *fundamental theories*. But that does not imply that it has been deposed also from the phenomena.

Not all readers were unreservedly delighted about the Minkowski space. We meet in the following P. W. Bridgman, who explains how the equations of the Lorentz transformation are interpreted as coordinate transformations in the four-dimensional Minkowski space and then mentions in a somewhat approving way the opinion of Herbert Dingle:

> A disaster of the first magnitude (from the point of view of the understanding, as distinct from the extension of the theory) occurred in 1908, when Minkowski transferred the subject from the realm of physics into that of mathematics.[30]

---

27 Minkowski, 'Space and Time,' 79–80.
28 Minkowski, 'Space and Time,' 81.
29 Minkowski, 'Space and Time,' 82–83.
30 Percy Williams Bridgeman, *A Sophisticate's Primer of Relativity* (London 1963), 10.

2.4. A Defence of Unique Space

We hinted above that the operationalist Percy Williams Bridgman was not quite at ease with the disappearance of place into a shadow in the four-dimensional world. He explains how geometric intervals are measured in the three-dimensional space. Many complicated operations can be used, but they all lead finally to an 'intuitive result':

> All the complicated processes receive their meaning and significance from the fact that there is in the background a single definite procedure to which we can return, and from which *we do not want* permanently to get away. If one asks 'why' it is that the interval is so fundamental for description, I think there is no answer. It is a brute fact, that every individual observer finds that the interval as determined by him with meter sticks stationary with respect to him is especially adapted to the description of nature. The fact that nature itself provides this unique method of description would seem to rob the contention that all frames of reference are equally significant of some of its intuitive appeal.[31]

One notes an oddity in Bridgman's text. On the one hand he says that every *individual observer* finds a certain procedure useful; on the other hand, he assures that *nature itself* has provided this unique procedure. If nature itself has provided it, it has provided it *to all observers* (at least the terrestrial ones) together; and as a consequence, they proceed in the same way. Bridgman next emphasizes still more the role of nature itself, saying that it offers us a unique system of coordinates, namely one fixed on the stellar universe, as Ernst Mach has noted. Bridgman thinks that Einstein does not find this reference system significant, because it does not fit with his confinement to 'fields'. According to that restriction, events can be significantly correlated only to their immediate vicinity. Somewhat incorrect, of course, is to speak of a unique *system of coordinates,* because it is always possible to apply a purely spatial coordinate transformation. Rather we should speak of a unique space. Mach was not the first to refer to the fixed stars for securing the *place* of things; that was done already by Copernicus.

Bridgman's ideas concerning a unique system of reference provided by nature itself have later been realized. Moreover, the system of reference can be determined in such an operational manner that presumably Bridgman himself would have been satisfied. We turn soon to the fact that Einstein himself had *de facto*

---

[31] Percy Williams Bridgman, 'Einstein's Theories and the Operational Point of View,' in: Paul Arthur Schlipp (ed.), *Albert Einstein: Philosopher-Scientist* (New York 1957), 313–332, 350–351.

already much earlier abandoned the assumption of the equality of all coordinate systems when he in his model of the universe divided space-time uniquely to space and time. That division gives us a space which is truly absolute; for if we accept equations (2) of the special relativity theory and signify a particular time coordinate as 'absolute', we automatically also signify a particular three-dimensional space as 'absolute'. We have mentioned earlier that feature of equations (2).

3. *Cosmological Perspectives*

3.1. *Einstein Abandons the Special Theory of Relativity*

Einstein had arrived in 1905 at the theory we sketched above. That theory entails the assumption of a *global* Lorentz-invariance, valid throughout the entire world. Einstein in his later reminiscences mentions his concern: could the principle of relativity in any way be extended to systems in accelerated motion? Einstein started already in 1907 on his road towards the fulfillment of this extended principle. Such fulfillment implied that the 'restricted' principle taken as the foundation of the special theory of relativity needed revaluation. Something was wrong with the Lorentz-invariance (and therefore with the universality of the equations (2)), if the new principle of relativity was feasible. The mathematical foundations of the new theory were laid using a modification of the four-dimensional 'world' constructed by Minkowski. The gravitational field in this modification was built into the geometric structure of the world. Einstein realized in august 1912 that the differential geometry developed first by Carl Friedrich Gauss, then by Bernhard Riemann and many later mathematicians, was the right mathematical weapon for a theory which later received the name *general theory of relativity*.[32] Einstein wrote in 1922 about his solution of the problem of gravitation:

> ...The problem remained insoluble to me until 1912, when I suddenly realized that Gauss's theory of surfaces holds the key for unlocking this mystery. I realized that Gauss's surface coordinates had a profound significance. However, I did not know at that time that Riemann had studied the foundations of geometry in an even more profound way ... I realized that the foundations of geometry have physical significance.[33]

---

32 Pais *'Subtle is the Lord...',* 209–210.
33 Einstein, *The Meaning of Relativity,* 211–212.

The combination of gravitation with geometry makes the interpretation of the fundamental concepts of the general theory of relativity somewhat difficult. The difficulties are aggravated when we pass from the locally 'almost Euclidian' situation to the grand scale of the cosmos or to strong gravitational fields. In the scientific biography of Einstein, the conceptual change from special to general relativity is passed over with slight attention:

> Its one kinematic novelty was perfectly transparent from the start: Lorentz invariance is deprived of its global validity but continues to play a central role as a local invariance.[34]

3.2. The Problems of Stationary and Expanding Universes

Einstein presented in 1917 a 'world model' for the global structure of the cosmos built upon his theory; we turn to its problems. *The Meaning of Relativity,* first published in 1922, is generally taken to be Einstein's own best and most complete exposition of both the special and in particular the general theory of relativity. In 1922 the possibility of an expanding universe was not yet an actual concept; therefore Einstein retained the concept of stationary world. For this reason he wrote about the four-dimensional universe, governed by the equations of general relativity theory:

> In this imagined universe all points with space direction will be geometrically equivalent; with respect to its space extension it will have a constant curvature, and will be cylindrical with the respect of its x_4 – co-ordinate [time coordinate]. The possibility seems to be particularly satisfying that the universe is spatially bounded and thus, in accordance with our assumption of the constancy of [the density] s, is of constant curvature, being either spherical or elliptical.[35]

Einstein has divided the four-dimensional space-time *globally* in space and time that have quite different properties. With respect to the spatial coordinates, the curvature of the universe is supposed to be constant; and Einstein obviously prefers a positive value for this constant. The time coordinate differs markedly from the spatial ones, and it makes the four-dimensional world 'cylindrical'. In the context of the citation above, Einstein *nowhere hints* at the fact that when he proposed such a structure of the universe he *contradicted his own principle of*

---

34 Pais *'Subtle is the Lord...',* 266.
35 Einstein, *The Meaning of Relativity,* 99.

relativity. Clearly, he does this when we accept his own practice of taking that principle to be relevant to the *material world itself* and not only the equations governing general laws.

Arthur Stanley Eddington wrote in 1920 an exposition of Einstein's theory of gravitation. He describes the early stationary relativistic cosmologies.[36] Differing from Einstein, Eddington clearly emphasizes the difference in the concepts of space and time in the special relativity theory and the cosmological models:

> Two objections to this theory may be urged. In the first place, absolute space and time are restored for phenomena on a cosmical scale. ... The world taken as a whole has one direction in which it is not curved; that direction gives a kind of absolute time distinct from space. Relativity is reduced to a local phenomenon; and although this is quite sufficient for the theory hitherto described, we are inclined to look at the limitation rather grudgingly. ...[37]

Einstein wrote in 1946 an Appendix to the third edition of *The Meaning of Relativity*, titled: *On the 'cosmological problem'*.[38] At that time the expansion of the universe was already known, and the expansion models had replaced the stationary models. In this context Einstein writes:

> We observe that the systems of stars, as seen by us, are spaced with approximately the same density in all directions. Thereby we are moved to the assumption that the *spatial* isotropy of the system would hold for all observers, for every place and every time an observer who is at rest as compared with the surrounding matter. On the other hand we no longer make the assumption that the average density of matter, for an observer who is at rest relative to neighbouring matter, is constant with respect to time. With this we drop the assumption that the expression of the metric field in independent of time.
>
> We now have to find a mathematical form for the condition that the universe *spatially speaking,* is isotropic everywhere. Through every point P of (four-dimensional) space there is a path of the particle (which in the following will be called 'geodesic' for short), ...
>
> ...One has to choose the radial directions as timelike and correspondingly the directions of the surfaces of the family as spacelike.[39]

The cosmologies of Alexander Friedmann and Abbé Lemaitre and the observations of Edwin Hubble and others have here brought a change to the earlier exposi-

---

36 Arthur Stanley Eddington, *Space, Time and Gravitation* (Cambridge 1966), 159–166.
37 Eddington, *Space, Time and Gravitation,* 162–163.
38 Einstein, *The Meaning of Relativity,* 104–126.
39 Einstein, *The Meaning of Relativity,* 107–108.

tion. Again, Einstein mentions, without any emphasis, the radical assumption: *the world itself* is uniquely divisible into space and time. It would scarcely have been possible to be more discrete about this innovation.

Accordingly, it was clear already from the first attempts at relativistic cosmology, that it was necessary to accept a three-dimensional space which at least in some sense could be characterized as 'absolute', and, consequently, also absolute time and absolute motion. In cosmology this idea has since been accepted as a matter of fact. For instance we read in a recent paper of Martin J. Rees:

> In homogeneous universes we can define a natural time coordinate, such that all parts of the universe are similar on the hypersurfaces corresponding to a given value of t.[40]

This statement implies that we can also define absolute space as the hypersurface mentioned by Rees. Similarly, we can define a stationary observer as one whose place agrees with that absolute space; all equality of observers has disappeared (Lorentz might rejoice had he heard about this). Next we may define absolute motion as motion relative to that observer (and here is the place for Newton to rejoice, pace Mach). All these definitions need something more than just a general theory about space and time; they need a universe with a spatial structure compatible with the *general theory* but not determined by it.

3.3. *The Microwave Background Radiation*

When philosophers of science speculating about relativity have drawn differences between 'observables' and 'non-observables', 'absolute motion' is often given as an example of the latter; it is often stated with some assurance that it cannot be disclosed using any mechanical or optical means. The 'principle of relativity', relying upon this assertion, includes the thesis that observers (reference systems) moving uniformly with respect to each other are 'equal' in the sense that no observable phenomenon can elevate any of them into a preferred position. The constant velocity of light was especially taken as proof of the equality. Light has, however, other properties than velocity: it demonstrates a Doppler-effect being *changed by the motion of the observer.* Observers moving relative to each other see different wavelengths of the arriving light. Supposing there were a good reason to view a particular wavelength as the *real one* and the others as distorted by the motion, this choice would put one observer into a preferred position. Nowadays microwave

---

[40] Martin J. Rees, 'Our Universe and Others,' in: Hermann Bondi and Miranda Weston-Smith (eds.), *The Universe Unfolding* (Oxford 1998), 51–68, 55.

background radiation has provided such a good reason; hence, it has weakened the dogma of the equality of observers. That weakening has rendered some of the 'philosophical' discussions about relativity somewhat old-fashioned. In the cosmological context this fact has been discussed since the discovery of background radiation; but, as far as the present author knows, such discussion has scarcely left any trace on philosophical speculation. This fact is somewhat astonishing, as the special position of some observers as 'stationary' had already become a possibility at the birth of the theories of the expanding universe, as we told above.

There are two 'golden moments' in modern cosmology. One was the discovery and interpretation of the red shift of galaxies by Edwin Hubble and others; the other was the discovery of microwave background radiation by Arno A. Penzias and Robert W. Wilson in 1965 in the Bell laboratory.[41] Both discoveries provided the real universe with a spatial and temporal structure which is not a consequence of any theory (at least a theory known at present). Cosmic background radiation is measured by the 'Planck spectrum'; in 1980 the corresponding temperature was measured at 2.96° K. (The value has, of course, been improved later.) The radiation was observed to be almost isotropic, with a small kinematic anisotropy due to the Earth's motion relative to the distant source of radiation. One of the investigators, Rainer Weiss, writes:

> The anisotropy, now definitely observed at a 10^{-3} level, is easily derived by a special relativistic calculation of the intensity measured by an observer moving relative to the walls of a blackbody cavity. The anisotropy retains a Planck spectrum but with an observation angle -dependent temperature (dipole term)...[42]

Hence, the effect can be mathematically treated using the special relativity theory; but the result may be interpreted as the discovery of a *factual* preferential role of one special system of reference, although such preference was not obvious in terms of the *theory* alone.

3.4. *The Measuring of Earth's Absolute Motion*

Already before the background radiation could in practice be used to measure Earth's absolute motion, the theoretical possibility of such a procedure was well understood. Steven Weinberg wrote in 1977:

---

41 Rees, 'Our Universe and Others,' 55.
42 Rainer Weiss, 'Measurement of the Cosmic Background Radiation,' *Annual Review of Astronomy and Astrophysics* 18 (1980): 489–535, 520.

The thing that makes the direction dependence of the microwave radiation background such a fascinating subject for study, is that the intensity of this radiation is not expected to be perfectly isotropic. ... In addition, there almost certainly is a small smooth variation of the radiation intensity around the whole sky, caused by the earth's motion through the universe. ... If, for example, we suppose that the earth is moving at a speed of 300 kilometer per second relative to the average matter of the universe, and hence relative to the radiation background, then the wavelength of the radiation coming from ahead or astern of the earth's motion should be decreased or increased respectively, by the ratio of 300 kilometers per second to the speed of light, or 0.1 per cent. Thus, the equivalent radiation temperature should vary smoothly with direction. ... For the last few years the best upper limit on any direction dependence of the equivalent radiation temperature has been just about 0.1 per cent, so we have been in the tantalizing position of being about but not quite able to measure the velocity of the earth through the universe. It may not be possible to settle this question until measurements can be made from satellites orbiting the earth...[43]

Weinberg here identifies Earth's 'absolute' motion as motion 'relative to the average matter of the universe'. Such an idea was taken up already earlier, but any resolution of the question was difficult when the investigation was restricted to Earth's motion relative to the visible matter of the galaxies. Now a new possibility had opened up: we can investigate Earth's motion through a universal homogeneous *radiation field*. The assumption behind this possibility was that the matter of the universe is, on the average, stationary in respect to that radiation field. Presumably no reason arises to worry about this assumption, as it may be more reasonable to define 'stationary place' using the average radiation as reference than the average matter of the universe.

The satellite *Cosmic Background Explorer,* shortened *COBE,* was launched in 1989. The cosmologist Malcolm Longair writes about the significance of its measurements for the determination of Earth's motion:

...At sensitivity level about one part in 1000 of the total intensity, there is a large scale anisotropy over the whole sky associated with the motion of the Earth through a frame of reference in which the radiation would be the same in all directions. This is no more than the result of Doppler effect due to the Earth's motion, and as a result the radiation is about one part in a thousand more intense in one direction and exactly the same amount less intense in

---

43 Steven Weinberg, *The First Three Minutes: A Modern View of the Origin of the Universe* (London 1983), 76–77.

the opposite direction. The intensity distribution has precise the expected dipolar distribution and it turns out that the Earth is moving at about 350 km s^{-1} with respect the frame of reference in which the radiation would be 100% isotropic.[44]

Background radiation and the 'dipole term' caused by the observer's motion is described in numerous astronomical books and articles. The absolute velocity of the Earth is already taken as an obvious matter of fact, and it is used to calculate velocities more difficult to disclose directly from observation, such as the velocity of our Galaxy. That velocity is found to be about 600 km s^{-1}. For some reason the astronomers writing about these developments seldom state explicitly that these results run counter to the time-honoured *relativity principle,* taken by physicists to be one of the corner stones of the mathematical *theory of relativity.* As already mentioned, the discrepancy arises only if the 'principle' is taken to govern not only *physical laws* but also all *physical phenomena.* Even after this caveat we might understand Eddington's feeling that 'we are inclined to look at the limitation rather grudgingly'. How are we to react to the discrepancy? Shall we accept the 'conspiracy theory of natural laws', according to which the entire concept of relativity is only a hoax of nature, which is unwilling to disclose 'absolute' entities? Shall we assume that the postulates which Einstein took as his starting point are only approximations in the sense that, *after all,* the motion of the Earth *does have some* slight effect on electromagnetic and optical phenomena? Whichever alternative we choose, it seems that we cannot avoid a dilemma.

Philosophers of science seem to avoid mentioning these cosmological developments. The transition from Lorentz to Einstein is praised as a transition from antiquated to modern views of space and time. Einstein explained the Lorentz-contraction epistemologically, referring to the impossibility of defining a universal simultaneity. Lorentz explained the same facts physically, referring to the change in molecular forces caused by the motion. Einstein's explanation is glorified as a deep revelation of the true world structure, while Lorentz's explanation is degraded to an *ad hoc* stratagem. Shall we retain these judgments now when ideas of absolute space and absolute motion can no longer be ignored as Newtonian monstrosities, which are, in Minkowski's words, 'doomed to fade away into mere shadows'?

---

44 Malcolm Longair, 'Modern Cosmology: A Critical Assessment,' in: Hermann Bondi and Miranda Weston-Smith (eds.), *The Universe Unfolding* (Oxford 1998), 275–314, 284–285.

3.5. Physicists' and Astronomers' Concept's of Space

The developments described above give occasion to wonder. Both the physicists and the astronomers accept the relativity theory at least in outline as an excellent theory. They seem, however, have different views about the concept of space implied by that theory. It seems that the difference is related to a question we have repeatedly mentioned, namely the question concerning the relation between a general theory and its concretization. We give a somewhat aggravated formulation of the difference.

According to the *physicist*, the real object of science consists in *general theories*. He regards the concrete *cosmos* as an interesting exercise in which such theories can be applied. Hence, the physicist accepts Minkowski's thesis about the disappearance of space and time as independent entities and the elevation of the four-dimensional world to the role of the proper physical reality.

According to the *astronomer*, the real object of science is the *cosmos* as we see it around us. He is interested in *general theories* only as instruments valuable for the understanding of the concrete inhabitants of the cosmos. Hence, the astronomer accepts the division of the world into uniquely definable space and time, for such are to be found in the cosmos.

Everyone can verify for himself the difference between the attitudes inherent in physics and astronomy. Just take a look at the text-books of these sciences and the descriptions of the grand lines in the development of the physical sciences as depicted by their practitioners. The text-books of physics advance in the order of *theories:* you learn first about mechanics, then about acoustics, optics, thermodynamics, electrodynamics, theory of relativity, quantum mechanics, nuclear physics, etc. The text-books of astronomy advance in the order of the *concrete inhabitants of the stellar world:* you learn first about the Earth and its Moon, then about planets, the Sun, stars, stellar clusters, galaxies, clusters of galaxies, and the entire universe and its origin in a big bang. Similarly, the physicist conceptualizes the great lines in the history of physical science as a growth of knowledge from Aristotle's physics through the mechanics of Newton, the electrodynamics of Maxwell, the relativity theory of Einstein, and the quantum mechanics of Bohr, Heisenberg and others to the present theories. The astronomer describes the way of knowledge as a path leading from the Earth-centered cosmos of Aristotle to the moving Earth of Copernicus, to the discovery of the Solar System, to Herschel's and other astronomer's ideas about the stellar world, to Hubble's and his collaborators' view of a universe filled with galaxies, and to Friedmann's, Lemaitre's, and the modern cosmologist's ideas about the expanding universe originating in a big bang, which somewhat later gave rise to the microwave radiation penetrating the entire universe.

Consequently, the physicist looks at *space* from the perspective of general theories, and the astronomer from the perspective of our unique cosmos. I wondered why the differences between these views have not often been clearly stated. Perhaps the reason lies in a reluctance to pinpoint discrepancies in such an excellent theory as the theory of relativity. For such discrepancies lead the physicists and astronomers in different directions to search for the reality of space and time.

12. Spinoza in Context

A Holistic Approach in Modern Terms[1]

Rainer E. Zimmermann (Munich and Kassel)

1. Introduction

Quantum gravity, as most promising candidate for a 'theory of everything' (TOE), sets out to describe the real world in the *philosophical* sense: as the foundation of the *modal* world we are able to perceive. Although utilizing most recent techniques of mathematics and theoretical physics, quantum gravity is thus reproducing the ancient quest for finding the answer to a very old problem which is intrinsic somehow in the very structure of human consciousness: 'What is the nature of the discrepancy between the world perceived (which is constituted by means of mapping the environment) and the world existing independent of human perceptions?' In other words: 'What is the precise difference between a picture (or model, or metaphor) of the world and the world as object of which this picture is made by the process of mapping?' Spinoza was the first philosopher who dealt with this problem in a modern language which is still accessible to us today, and it was he who initialized a whole line of philosophizing in following his basic ideas, – via Fichte, Hoelderlin, Schelling, and Hegel up to modern French existentialism and structuralism, and up to the materialist philosophy of Ernst Bloch. The important point is that it is also Spinoza himself already who develops in a nutshell all relevant consequences of his ideas leading forward to a realistic materialism rather than to an idealistic sort of philosophy. And in exactly this sense, his ideas can be visualized as forerunners of those ideas which are underlying the philosophical motivation of recent research.

---

1 In section 2, I follow the general line of argument presented in the preprint version of an earlier paper: 'Loops and Knots as Topoi of Substance. Spinoza Revisited.' (Presently submitted in a revised version to http://www.arXiv.org/ – gr-qc/0004077 (v2).) In section 3, I also paraphrase some main results of this paper.

2. Spinoza in Context

Spinoza does not only cumulate the various strings of the philosophy of his time. He also visualizes philosophy as something which is practically identical with ethics. For him, philosophy is a theory of the conditions according to which human life is being defined. And this definition can only be rational, if it is succeeding with respect to an ethical frame of references. Ethics itself, unfolds the conditions according to which the human striving can be realized. Humans are capable in this sense, of finding and conserving their own mode of being, if they act according to adequate knowledge. Hence, there is a close connection between freedom and insight. It is necessary therefore, to find adequate ways (inveniri) in an appropriate project which is to be designed by humans themselves. If Spinoza's *Ethics* (1p34)[2] states that the power of God is his own *essence* (Dei potentia est ipsa ipsius essentia), then, for humans, one could add in an existential sense that the power of humans is their own *existence* (Humani potentia est ipsa ipsius existentia). And the adequate form of this existence is being prescribed in terms of the virtue which leads forward to blissful happiness. For humans therefore, virtue in this sense and power, are identical. (4d8)[3] Not only is this a mere re-formulation of stoic ideas, but Spinoza also tries to define substance (God = nature) as *causa immanens* of the world, but in terms of a twofold perspective taken according to whether the relationship between God and world is visualized under the substantial (real) perspective of God himself, or under the modal perspective of humans who represent a finite mode of what worldly exists. Obviously, the former perspective can be taken by humans in speculative terms only, but because God represents himself completely in each of his attributes, it is possible that humans can eventually grasp his (substance's) existence in principle, provided they have developed the adequate knowledge about this due to their adequate reflexion. In this sense, everything is in God (quicquid est, in Deo est), but the viceversa is also true.

On the other hand, reflexion itself is an outcome of the organization of substance: The constitution of the latter according to which it is productive with

---

2 Throughout the text I use the English notation when referring to Spinoza's *Ethics*, the first number giving the part, the letter referring to the state of the proposition, in a somewhat self-explanatory manner, the second number following the usual listing of propositions.

3 Although, for reasons of simplicity, we keep 'power' here for Spinoza's 'potentia', it should be noted that 'potential' (in the sense of capacity, ability, ...), contrary to 'potestas' (power) is certainly a better translation. In so far I follow the opinion of the English translator of Negri's book on Spinoza: *The Savage Anomaly* (L'anomalia selvaggia: Saggio su potere e potenza in Baruch Spinoza, 1981) (Minneapolis 1991). In Italian, there is a natural difference between the two expressed in the title (potere, potenza). The same is true for French (pouvoir, puissance). I claim this difference also for German (Macht, Vermögen). See the translator's remarks in Negri's book, xi sqq.

respect to the field of modi, is its organization in terms of attributes.⁴ This means that God does not really produce attributes, but he (as substance) is *attributively organized* instead. Hence, God is *causa sui* only in so far as he produces everything what there is, but this is only true with respect to the finite perspective of humans. Nevertheless, a finite mode is in God, because it is a created mode within the totality of nature. But in this mode as its cause, it is only God who acts as immanent cause of permanency, it is not the totality of nature which is acting as this cause.⁵ Hence, the central position of nature which is classified by Spinoza in a twofold way according to the classification given earlier by Averroes: He actually differs between *natura naturans* (the actively creating nature producing things) and *natura naturata* (the passively produced nature which is the outcome of processes performed by natura naturans). The former represents the productivity of God, the latter its result. *Nature is the form of mediation in which God acts upon the world as seen (and interpreted) under the modal perspective of humans.* But in reality, he does not think himself, because it is only the humans who do. (2a2) And therefore, he is not a spirit either.⁶ So nature has an important role to play within the frame of references which constitutes the world. But as it is only humans that think, the basic concept of worldly orientation is *intersubjectivity* in the first place, nature only in second. It is *reflexion,* and it is the *political form of communication* that determine adequate knowledge. But the latter can only be achieved, if the structure of the world in terms of its nature is uncovered and logically displayed. Hence, to study nature means to lay the ground for adequate knowledge, and in the end, for adequate *action* according to ethical principles. And this is a holistic viewpoint indeed.

Spinoza seeks to derive his ethical theory from this basic understanding of nature in the first place, the psychology of humans showing up within this framework as *human nature,* a particular case of nature in general. In this sense, every human action must be conceived as a manifestation of nature.⁷

Note that this is the celebrated aspect of determinism in Spinoza, though not as we know it: Instead of 'determinism' it would be more adequate to speak of 'necessitarianism', because nature is actually determining behaviour, but in an intrinsically contingent way. Hence, there is always a sufficient reason for everything which happens, and a kind of causal closure (consistency), but there is not necessarily a fixed 'program' of processes. (We actually would argue today that physics is determining all processes in the world, but very much in terms of a basic

---

4   For a precise discussion of this aspect see W. Bartuschat, *Spinozas Theorie des Menschen* (Hamburg 1992), 66.
5   Bartuschat, *Spinozas Theorie des Menschen,* 37, 44–49 par.
6   Bartuschat, *Spinozas Theorie des Menschen,* 65.
7   D. Garrett, 'Spinoza's ethical theory,' in: D. Garrett (ed.), *The Cambridge Companion to Spinoza* (Cambridge 1996), 267–314, 270.

framework, of some arena for processes. So having found out that in the end, 'everything is physics', is certainly a true result and statement, but it is also completely unsatisfactory without having done the hard work – namely re-constructing the various levels of the hierarchy of complexity and following up the lines of mediation towards the phenomena actually being observed.)

This has a decisive consequence for *praxis* visualized as human existence: For Spinoza, the striving for conserving the own being (conatus in suo esse perseverandi), this characteristic kind of *eigen-being,* can be interpreted in terms of an ethics which unfolds the conditions under which this striving can actually be realized. Under them, therefore, the individual human being can indeed arrive at its own being (eigen-being).[8] Provided he/she applies what Spinoza calls the *true method* consisting of 'the knowledge alone of the pure understanding, of its nature and of its laws. To acquire this method, you must first of all distinguish between intellect and imagination, or between true ideas and the others, that is fictitious, / false and doubtful ideas, and, absolutely speaking, all those that depend on memory alone.'[9]

Hence, this is the kind of 'ideal of a unified science'[10] Spinoza is aiming at. In principle, he seeks primarily to improve the character of human beings by improving their self-understanding. He sets out to show how humans can achieve a way of life that largely transcends the mere transitory desires leading to an autonomous control over passions.[11] It is in this way that Spinoza's monistic and naturalistic system speaks 'most cogently and persuasively to the twentieth century.'[12] In fact, it is doing so by explicating three basic aspects which have become very important for modern research in recent times: 1) visualizing a transcendental materialism with physics 'at the bottom', 2) in a radical approach to interdisciplinarity, and 3) aiming at mediations leading up to ethics and politics.

All of this is very much in the sense of what we would expect today of a close interrelationship among philosophy and the sciences and arts. And this is what Negri actually means when speaking of Spinoza as 'anomaly': 'Spinoza is the anomaly. The fact that Spinoza, atheist and damned, does not end up behind bars or burned at the stake, ... can only mean that his metaphysics effectively

---

8   Bartuschat, *Spinozas Theorie des Menschen*, x par. – See also for an unfolding of intersubjectivity based on this ethics: 186 (referring to 4p18) and 200. With strong Sartrean connotations actually, because such a human can be easily visualized as one 'who is able to become what he/she actually is'.
9   As quoted by A. Gabbey, 'Spinoza's natural science and methodology,' in: D. Garrett (ed.), *The Cambridge Companion to Spinoza,* 142–191, 171sq. – referring to the epistle (37) to Bouwmeester.
10 E. Curley, *Behind the Geometrical Method: A Reading of Spinoza's Ethics* (Princeton 1988), title of ch.1 sect. 2, p. 4; p. 6.
11 Garrett, 'Spinoza's ethical theory,' 267sq., par.
12 D. Garrett in his 'Introduction' to *The Cambridge Companion to Spinoza,* 1–12, 2.

represents the pole of an antagonistic relationship of force that is already solidly established: The development of productive forces and relations of production in seventeenth-century Holland already comprehends the tendency toward an antagonistic future.'[13]

In his striving for an *experientia sive praxis*[14], Spinoza aims at a theory of the world, beginning with physics, ending up with social systems, proposing educational means of eventually approaching a state which, in an equilibrium of the individual and the institution, harmonizes human spirit in order to let it participate in the larger harmony which is expressed in the material attribute of godly substance. This is obviously the reason for the recent reception of Spinoza's in explicitly Marxist terms (notably in Althusser). It is Tosel who recognizes an *ethico-political* consequence emerging from Spinoza's approach, basing it on definite materialistic *philosophemes* and a collection of motives generic for Marxist theory: 'L'onto-théologie est éthico-politique: Dieu, c'est-à-dire la légalité d'une Nature immanente, se traduit humainement dans l'immanence d'une société rationellement réglée d'hommes capables de penser et agir. Et l'éthico-politique est à son tour ontologique: l'homme libre est une possibilité de la nature anonyme.'[15] In this sense, Tosel shows that for Spinoza, all of reality emerges from another reality which is basically material, the unity of which is immanent in the variety, and coincides with its own space of productivity. Hence, cosmology, physics, and logic are the basic fields for constituting anthropology.[16] The results following immediately from this are exactly the topics which are our interest today when trying to find a new synthesis of philosophy, science, arts, and praxis (of which the first three are a part). And we can realize the strong existentialistic connotation being handed down to us from Spinoza: Because of the elements just mentioned, philosophy has a fundamental interest in liberating humans, in aiming to a positive liberty of individuals against all kinds of heteronomy. Philosophy is theory and praxis of autonomy. And in being a science of humans, philosophy is also an instrument of critique, especially of critisizing all authorities which do not base their principles on communicative arguments. Research about life is therefore research of (and about) *free* life in the first place, of a life which has no other generic principle than its physico-psychic unity. For this, nature is the original reality, and as such nature is intelligible, and it is anterior to all thinking.[17]

In order to clarify the concept of substance somewhat, it might be the best to look for its relation to the fundamental categories of space and time in more detail:

---

13 Negri, *The Savage Anomaly*, xvii (preface).
14 Negri, *The Savage Anomaly*, 183 (title of section).
15 A. Tosel, *Du Matérialisme de Spinoza*, (Paris 1994), 18.
16 A. Tosel, *Du Matérialisme de Spinoza*, 133, 135, par.
17 A. Tosel, *Du Matérialisme de Spinoza*, 187sq., p. 190, par.

Essentially, the attribute of extension can be visualized as space itself.[18] The actual difference between Leibniz and Spinoza can be phrased then in terms of the relationship between objects and space. The former explains away the region and stays with the relations among bodies, taking the region as an alternative way of expressing facts about them. The latter explains away the objects and stays with the region. In referring to a famous example, Bennett formulates: 'If there is (...) a pebble in region R, what makes this true is the fact that R is *pebbly* (which) stands for a certain monadic property that a spatial region / can have. If the pebble moves (...), what makes this true is the fact that there is a continuous change in which regions are pebbly: The so-called movement of a pebble through space is like the so-called movement of a panic through a crowd.'[19] Note that this gives a relatively clear explanation of how we should visualize *motion-in-itself*. In this sense, time does not remain as a fundamental category. Spinoza states in his 12th letter that '... *tempus* (is) nothing but a mode of the imagination which ought to mean that in a true fundamental account of the whole of reality th(is) concept ... would not be used. (Hence,) all our measures – of time and space and of things spatial and temporal – are superficial and "imaginative", and not part of the basic, objective story.'[20]

Although this concept is widely agreed upon, at the same time, it has been the target of many discussions. It can be shown however that Spinoza ties time to the imagination (2p44s), basically referring it to human perceptions of varying motions of bodies.[21] Hence, temporality actually emerges in the transition from infinity to finite modes (of thinking) as laid down in 2p8 and its corollary.[22] In fact, as Bartuschat mentions explicitly in his discussion at this point, the corollary has the important function of marking the transition to finite modes, and it is here where Spinoza actually changes the perspective of description in his *Ethics*.[23]

---

18 See e.g. J. Bennett, 'Spinoza's Metaphysics,' in: D. Garrett (ed.), *The Cambridge Companion to Spinoza*, 61–88, 69.
19 J. Bennett, 'Spinoza's Metaphysics,' 70sq. Cf. 70.
20 J. Bennett, 'Spinoza's Metaphysics,' 77, referring to epistle 12 and to 1p15s.
21 M. D. Wilson, 'Spinoza's theory of knowledge,' in: D. Garrett (ed.), *The Cambridge Companion to Spinoza*, 89–141, 139 n.51.
22 Bartuschat, *Spinozas Theorie des Menschen*, 85.
23 See in more detail, Bartuschat, *Spinozas Theorie des Menschen*, 87. – Also J. D. van Zandt agrees to this when stating that '(d)uration, then, is solely to be found in Natura naturata ...' 'Res extensa and Space-Time Continuum,' in: M. Grene and D. Nails (eds.), *Spinoza and the Sciences* (Dordrecht 1986), 249–266, 256. But it is not clear why he then relates this result to considering the space-time manifold 'as a whole' to 'eternal natura naturans'. (van Zandt, 'Res extensa and Space-Time Continuum', 257, 259). – Ironically, when Kouznetsov once referred to quantum theory as representing natura naturans, he was not so far off as Paty thinks in the article following van Zandt's, cf. M. Paty, 'Einstein and Spinoza', M. Grene and D. Nails (eds.), *Spinoza and the Sciences*, 267–302, with a view to B. Kouznetsov, 'Spinoza and Einstein (!),' *Revue de Synthèse* 88 (1967), 45–46; fasc. 3, 31–52.

Obviously, if temporality is part of the human world view in modal terms, then there is no problem of determinism. But still, the question is as to the actual structure of the human mode of being which is also the human mode of perception. Of course, Spinoza's identity theorem (2p7) is the main topic around which this question centres. If 'thought is co-extensive with materiality'[24], then the ethical implications of Spinoza rest basically on the fact that human thinking is not a mere passive reflexion of the body's vicissitudes: 'As well as being mapped onto a segment of the material world, the mind is inserted into the totality of thought.'[25] Hence, a singular individual is not a self which also strives for something, but a singular individual is a self-in-striving, and it has a *project structure* (very much in the sense of modern existential philosophy). But this project which in the end shows up as the dynamical core of existence, is a representation of the human's finite mode of being, which points toward the adequate form of knowledge humans can have, because it is this knowledge in which they actually conserve their own being. And the generic form of realizing existence is *acting* according to the results of adequate reflexion.[26]

Hence, human reflexion means nothing else but explicating the symbolic traces in existence which intrinsically point to the true mediation of what there is with substance, visualized in terms of the latter's attributes. Modelling the world in human terms means modelling the configuration of (worldly) attributes. Obviously, the latter cannot be modelled without modelling its foundation (substance) at the same time. This is the point where *sceptical* philosophy is clearly based and dependent on *speculative* philosophy. In fact, human reflexion has the actual task of organizing the infinite[27] by means of the finite. And both of them are 'knotted to each other' in terms of the consequences of the identity theorem 2p7: 'Mental items can be mapped onto bodily items in a way that preserves causal connectedness.'[28] Hence, if M is the category of mental states, and B is the category of bodily states, respectively, then there is a functor $\varphi: B \to M$ (preserving identities and compositions). If so, probably, we could expect that the diagram

$$\begin{array}{ccc} M_1 & \to & M_2 \\ \uparrow & & \uparrow \\ B_1 & \to & B_2 \end{array}$$

commutes, for any pair of bodily states, and mental states, respectively, when the horizontally parallel arrows $\to$ refer to morphisms in the respective categories,

---

24 Wilson, 'Spinoza's Theory of Knowledge,' 115.
25 G. Lloyd, *Part of Nature. Self-Knowledge in Spinoza's Ethics* (Ithaca 1994), 27.
26 Bartuschat, *Spinozas Theorie des Menschen*, xi, 132sq.
27 Negri, *The Savage Anomaly*, 52.
28 This is an alternative formulation by Bennett, 'Spinoza's Metaphysics,' 78.

and the vertically parallel arrows ↑ define generic 'liftings'. (In fact, one would expect appropriate isomorphies to hold here.)[29] These relationships secure that humans are capable at all of adequately interpreting the symbolic traces intrinsic to the world they observe. And this is the reason why speculative philosophy is not only a necessary action to be undertaken, but also one which might result in sufficient insight.

For Spinoza therefore, substance is what is in-itself and what can only be comprehended by itself and out of itself. This means that it is something whose concept does not need the concept of anything else in order to be formed. Hence, substance is its own reason (causa sui), and its essence involves its own existence. But humans, being a finite mode of this infinite substance, can only perceive their world in terms of attributes of this substance, but not substance itself. In fact, their world is actually being defined in terms of these attributes of which there are infinitely many, of which humans however can only perceive the two which fall into their mode of being: matter (res extensa) and mind (res cogitans). Note that not only is the world not the same as substance (hence all that there is for humans is only one special aspect of all that there is in reality), but humans are also permanently modelling the attributes, in a kind of recursive approximation, rather than perceiving them as they actually are. Hence, the relationship between the real perspective of substance and the modal perspective of humans is difficult to visualize: E.g., although the infinite substance is undivisable (has no parts), humans actually perceive parts of attributes, but this difference is only given in modal, not in real terms. (1p15s)

---

29 That is we would expect that there are generic projections p and q such that the compositions with the respective liftings, a and b say, reproduce the various identities: p o a = 1 (M), a o p = 1 (B), and the appropriate for q and b. – The version of the theorem which Bennett gives is not very clear when starting with the idea of parallelism from the outset (as he actually does). He says that mode identity would mean that 'if M is correlated with B under parallelism, then M = B.' (Bennett, 'Spinoza's Metaphysics,' 79) In fact, 'identity' in the theorem 2p7 refers to the 'order and connection' of ideas on the one hand, and of things on the other. ('Ordo, & connexio idearum idem est, ac ordo, & connexio rerum.') It does not mean that ideas and things are actually the same. It is true however that 'the role of the attributes is to combine with the transattribute modes to get the latter into a form in which we can think them.' Indeed, the 'attributes let the / modes come through. It is as though the modes were words written in a script to which intellect is blind, and the attributes make the message of the modes accessible to intellect by reading them aloud, expressing them.' (Bennett, 'Spinoza's Metaphysics,' 87sq.) This relates to the aspect of self-narration which I have discussed at another place. See my 'Prosperos Buch oder Echolot der Materie. Zum hypothetischen Natursubjekt bei Ernst Bloch: Bilanz & Ausblick,' *VorSchein*, 15 (Nürnberg 1996), 40–57. Reflecting about this there is the possibility of discussing it in terms of relating modes to (mathematical) logoi – in the sense of René Thom –, and attributes to (mathematical) topoi, respectively. I have discussed this in more detail in my 'Emergenz und exakte Narration des Welthaften: Zur Naturdialektik aus heutiger Sicht,' *System & Struktur* III/1, 1995, 139–169. See also my forthcoming book on 'The Physics of Logic' (2000).

Hence, for humans, substance is the foundation of being, and as such it is *non-being*. It is not nothing, because the world, as visualized in modal terms, has eventually emerged from substance. In a way, we can say that the world has been produced by substance out of a field of possibilities. Hence, worldly objects come into existence by some initial emergence of the world which is thus an exteriorization of substance (in the sense that substance unfolds its organizational structure – which is not really a process when visualized under the real perspective of substance, but rather an equivalent self-representation of substance itself). Note that when we as humans, are talking about this emergence, the very language we have to use points to a dynamical process of some kind. The reason for this is that we have to think according to our usual (modal) terms of reflexion which involve temporality one way or the other. But, in reality, substance within this picture, is not actually triggering dynamical processes, it is already structured in a way that it is differentiating itself in terms of infinitely many self-representations, of which one is what we interpret as our world. Hence, motion in terms of substance is ill-defined. It is rather that substance is constituted as potentially self-moving in the sense that it is in a state of permanent self-fluctuation which represents an intrinsic sort of motion, an abstract motion (a motion in-itself), and a potential for concrete, worldly motion, at the same time. It is in this way that the intrinsically dynamical constitution of substance points to a concept of freedom which basically means 'freedom to eventually produce a structured world'. This could be interpreted as a kind of absolute freedom which in itself can be taken as a practical definition of substance.[30]

The world is therefore a kind of deficient state of substance as it appears only in terms of restrictions (as being primarily finite). Substance however, as foundation of being, is itself without foundation. Hence, it is constituted in terms of self-reference, and it propagates aspects of this self-reference into the world. In this sense, substance is beyond space and time, it is basically non-local, it may be visualized as 'pre-geometry'. At the same time, the world as the product of substance is constituted in a transcendental sense, because there is an immanent tendency of the worldly towards returning to its own origin. (Under the real perspective of substance this means that substance has itself the tendency to re-integrate its own unfolding (representation) into itself, which is in fact nothing but an alternative expression of its own totality. In principle then, the world can be re-interiorized again, and it is this final stage of development, in which worldly existence is sublated again in its original (primordial) unity – very much in the threefold Hegelian sense.)

---

30 I have discussed this aspect in my 'Freiheit als Substanz: Metaphysische Aspekte von Initialemergenz und kosmischer Evolution,' in: W. Saltzer, P. L. Eisenhardt, D. Kurth, R. E. Zimmermann (eds.), *Die Erfindung des Universums? Neue Überlegungen zur philosophischen Kosmologie* (Frankfurt a.M. 1997), 45–71.

3. Pre-Geometry

The close relationship of the preceeding line of argument in Spinozist terms is not completely straightforward when visualized with a view to possible holistic approaches of philosophy today, in really 'modern terms' so to speak. However, owing to the conceptual preparations undertaken in the first part of this paper, it may become sufficiently accessible and acceptable to eventually re-construct the formal framework at least, of this original approach, referring back to the three basic aspects of a possible interface between Spinozist metaphysics and modern philosophy. Hence, again, we visualize a theory, which shows up as a 'transcendental materialism with physics "at the bottom", in a radical approach to interdisciplinarity, and aiming at mediations leading up to ethics and politics.'

What is then today the respective 'physics at the bottom'? This turns out to be what we call *quantum gravity* which is one of the most promising candidates for a theory of everything (TOE) in modern physics. In fact, there are two such theories of quantum gravity which are heavily competing with each other, the theory of superstrings, and loop quantum gravity, but we will concentrate here on the latter, for reasons I have given elsewhere.[31]

Let us note first that since the advent of general relativity theory, the strive for a unification of science (philosophically speaking: for grasping the foundations of physics and hence the foundation of the world) has entered a new phase of intensity. For Einstein himself, it appeared appropriate to look for a unification in terms of gravitation and electromagnetism, the two fundamental forces (or interactions) at the time. To be more precise: Einstein's idea was to actually include electromagnetic fields in his gravitational field equations by introducing them (by means of their potentials) into the metric components of space-time, in the first place. As Wolfram Voelcker, Can Yurtoeven, and myself have shown at another place[32], this can be interpreted in terms of *visualizing substance as spacetime geometry* in the sense of general relativity, representing gravitation, at the same time. (Indeed, as Lewis Feuer has pointed out earlier[33], it is quite certain that Einstein developed his theory under the impression of discussing Spinoza's philosophy when in Prague.) The subsequent advent of quantum theory however, to which Einstein himself contributed some fundamental insight, led to a much more complex situation, because not only did two more forms of interaction (weak and strong fields) turn up, but the basic interpretations of relativity on the one side and quantum theory on the other, differed according to their respective (worldly)

---

31 Cf. my 'Loops and Knots as Topoi of Substance'.
32 'Philosophical Aspects of Spin Networks. An Alternative Einstein Memorial,' ANPA 21, Cambridge (UK), 1999. To be published.
33 L. S. Feuer, *Einstein and the Generations of Science* (New Brunswick 1982).

domains: on the one hand, relativity being defined in terms of a four-dimensional space-time manifold with a Lorentz metric (of signature -2), quantum theory on the other, showing up as being defined in terms of a high-dimensional Hilbert manifold with a positive-definite (Euklidean) metric. Hence, one ended up with two disjoint domains claiming to explain the same universe. So, although in the meantime various intermediate successes have been celebrated (such as the unification of the electroweak force, or the approach towards a *grand unification* of the latter with the strong forces: GUT), the ultimate goal, the unification of all forces (and matter) with gravitation in a *theory of everything* (TOE) has not yet been achieved.

During the last three decades, a basically different approach to unification has been put forward going back to an old idea of John Wheeler's: If it is not possible to unify the competitors within the world, it might be possible to unify them *outside* the world, the basic idea being to introduce an abstract mathematical structure from which space and time (and matter) as fundamental categories of the world could be eventually *derived*. It is quite straightforward in fact, to notice the connotation of substance here: If we define our world in terms of fundamental categories such as space and time (and matter), then everything outside the world from which we might be able to derive these fundamental categories is the *foundation* of the world and as such it is *non-being*. Hence, the idea is to visualize the world as a variety which has become out of a primordial (actually pre-worldly) unity. If so, then the next step, namely to formulate this the other way round: that the world is in fact this primordial unity as being observed as a becoming variety by its members who have restricted means of perception, is relatively small. Contrary to what Einstein thought, space-time-matter would not be substance itself, but only the latter's worldly attribute. And what is 'before' (and external to) the world, *pre-geometry*, would gain the connotation of a substance.

The question is how to reasonably approach such a conception in more detail. Because, technically, this would point to achieving a unified context for both gravitation and quantum theory anyway. A useful motivation for 'quantum gravity' of that sort has been given by John Baez recently[34]: There are three fundamental length scales, defined in terms of three physical constants coming from both regions in question ($\hbar$, the Planck constant divided by 2π, Newton's gravitational constant G, and the velocity of light c, respectively), which are important for relativity and quantum theory, at the same time. One is the Planck length $l_p = (\hbar G/c^3)^{1/2}$, of roughly the order of magnitude of 10^{-35} m. For length scales smaller than this, quantum gravity would be the appropriate tool to describe the physics there. Another one is the *Compton wavelength*, $l_c = \hbar/mc$, which basically indicates that

---

34 J. C. Baez, 'Higher-Dimensional Algebra and Planck-Scale Physics,' in: N. Huggett, Callender (eds.), *Physics Meets Philosophy at the Planck Scale,* (Cambridge 1999), in press. (I am quoting from the manuscript version of 28th January 1999, 4.)

the measuring of the position of a particle of mass m precisely within one Compton length, requires energy which is able to create another particle of the same mass. Hence, this length scale is characteristic for effects in quantum field theory. Finally, there is the *Schwarzschild radius* which basically defines the horizon of a black hole which has been formed by a collapsing star of mass m: $l_s = Gm/c^2$. (All constants up to numerical factors.) If m is now the Planck mass itself ($m_p = (\hbar c/G)^{1/2}$), then $l_c = l_s = l_p$. Hence, we would suppose that at the Planck scale, both domains of physics should show up in a somewhat unified way.

Thinking of the traditional division of classical relativity, and quantum theory, respectively, it is quite natural to ask whether the continuum picture of space-time is only an approximation which inevitably breaks down when approaching the Planck scale. And if so, whether there are constituents of space and time which show up according to some scheme of quantization, such as to construct quantum operators with discrete eigen-values. The microscopic structure of space and time would be determined then by eigen-values and eigen-vectors of purely geometric operators, and the macroscopic superposition of these would show up as the well-known space-time continuum (as a limit for large values). But this would also mean to abandon any underlying space-time structure we have got accustomed with (both in relativity *and in quantum theory* which is also based on a classical space-time background). As Ashtekar and Krasnov have pointed out, '... to probe the nature of quantum geometry, one should not begin by *assuming* the validity of the continuum picture; the quantum theory itself has to tell us, if the picture is adequate ...'[35] For their approach, referred to as 'loop quantum gravity', it is the goal therefore, to find a background-free quantum theory with local degrees of freedom propagating causally.

Essentially, Ashtekar and Krasnov refer back to Wheeler's conception, in actually introducing a split of space and time such that what changes in general relativity in dynamical terms is not the 4-distances within space-time, but rather the 3-distances within spaces as being nested in space-time. Hence, the dynamics is essentially one of 3-dimensional Riemannian spaces. This idea going back to John Wheeler in the sixties is discussed in Julian Barbour's new book[36], where Barbour points out that the 'key geometric property of space-times that satisfy Einstein's equations reflects an underlying principle of best matching built into the

---

35 A. Ashtekar, K. Krasnov, 'Quantum Geometry and Black Holes,' gr-qc 9804039 v2 (4/2/99), 4.
36 J. Barbour, *The End of Time. The Next Revolution in Our Understanding of the Universe* (London 1999). – Note that for the Einstein vacuum equations $R_{ab} = 0$, with space-time as M = R x S, where S is the (t = 0)-slice of M, $R_{0b} = 0$ are the constraints on the initial data, and the remaining equations give the evolution in time. The physical states are given as the subspace of diffeomorphism invariant states that are annihilated by the constraint corresponding to R_{00}. The equations expressing this fact are the Wheeler-deWitt equations, which, in the book of Barbour's, take a central position therefore.

foundations of the theory.'[37] The time separation of spatial slices shows up here as what Barbour calls a *distinguished simplifier*, as an ordering principle for making unfoldings simple.[38] If time is being visualized as a mere ordering principle, then, in philosophical terms, we are left with space as an attribute. Note however, that the *dimensionality* of space is only a finite representation then, which does not reflect the true nature of space, but only our modal attitude towards it with a view to spatial ordering.

The Ashtekar ansatz is also invariant under local SU(2) gauge transformations, three-dimensional diffeomorphisms of the manifold on which the fields are defined, as well as under (coordinate) time translations generated by the Hamiltonian constraint. The full dynamical content of general relativity is also captured by the three constraints that generate these gauge invariances.[39] So what we can do now is to compare the configuration variable of general relativity as known from gauge theories with the SU(2) connection A on a spatial 3-manifold, and the canonically conjugate momentum E with the Yang-Mills 'electric' field. Physically, the latter is essentially the triad and carries all information about space. This is where in quantum theory, the gauge invariant *Wilson loop functionals* are coming in: They are the path-ordered exponentials of the connections around closed loops. Hence, the name for the theory (loop quantum gravity). We will come to that in a moment.

Note however another relationship first: Usually, the kinematics of quantum theory is defined by an algebra of operators on a Hilbert space. The outcome of the physical theory will depend on the connection which can be uncovered between operators and classical variables, and on the interpretation of the Hilbert space as a space of quantum states. The dynamics is determined by a Hamiltonian (in general relativity called quantum constraint) constructed from the operators. The idea is now to express quantum states in terms of the amplitude of the connection, given by some functional of the type $\Psi(A)$ in the sense of Schrödinger. Functionals of this kind form a space which can be made into a Hilbert space provided a suitable inner product is being defined. Having done this, we can express the amplitude as $\Psi(A) = \langle A | \Psi \rangle$ in Dirac notation. Is now H our Hilbert space, then we define H_0 as its subspace formed by states invariant under SU(2) gauge transformations. Then it can be shown that an orthonormal basis in H_0 is actually a *spin network* basis.

What is a spin network? Basically, spin networks are graphs with three-valued nodes and spin values on the edges, their states being denoted by $|\Gamma\rangle$. To take the

---

37 Barbour, *The End of Time*, 176.
38 Barbour, *The End of Time*, 180.
39 C. Rovelli, 'Loop Quantum Gravity, Living Reviews in Relativity,' www.livingreviews.org. MPI Potsdam, 19.

norm refer to the mirror image of a graph and tie up the ends, forming a closed spin network $\Gamma\#\Gamma^*$ of value V such that

$$<\Gamma|\Gamma> = V(\Gamma\#\Gamma^*)$$

with

$$V(...) = \Pi(1/j!)\Sigma\varepsilon(-2)^N,$$

here j being the edge label, N the number of closed loops, and ε referring to the intertwining operation taking care of permutations of signs. The product is to be taken over all edges, the sum over all routings. Hence, the networks can be visualized in diagrammatic form such as to represent the underlying 'spin dynamics'. For instance, a diagram with vertices a, b, c, and i, j, k within the region of intertwiners such that $i+j=a$, $i+k=b$, $j+k=c$, can be interpreted as two particles with spin a and b which produce (create) a particle with spin c. Spin interactions of this kind can lead to the creation of new structures: Take a large part of the network (effectively representing a part of the universe) and detach from this small m-units, and n-units, respectively, as 'free ends'. The outcome of their tying up to form a new structure can be estimated in terms of a probability for the latter having spin number P, say. This turns out as being basically the quotient of the norm of the closed network and the norm of the network with free ends (times some intertwining operations). The *spin geometry theorem* tells us then that when repeating this procedure and getting the same outcome, then the new quotient is proportional to (1/2) cos θ, where the angle is one which is taken between the axes of the large units. Hence, it is possible to show that angles obtained in this way satisfy the well-known laws of Euklidean geometry. Or, in other words: This purely combinatorial procedure can be used to actually approximate space from a *pre-spatial* structure which is more basic. This idea has been due to Penrose who originally tried to base his concept of *twistors* on spin networks. In looking for primary concepts of an abstract structure from which space (and eventually space-time) could be approximated, starting from purely combinatorial elements, he began with looking more closely to the implications of angular momentum as to the re-construction of space.[40] The consequences of this approach have remained relevant until today, although twistors are not in fashion nowadays.[41]

---

40 R. Penrose, 'Angular Momentum: An Approach to Combinatorial Space-Time,' in: T.Bastin (ed.), *Quantum Theory and Beyond* (Cambridge 1970), 151–180.
41 See however S. A. Huggett et al *The Geometric Universe, Science, Geometry, and the Work of Roger Penrose* (Oxford 1998). – For an interpretation of twistors within the framework of substance metaphysics see my 'Initiale Emergenz und kosmische Evolution: Zur Rekonstruktion

A generalization of spin networks and a connection with knot theory has been achieved more recently by Carlo Rovelli and Lee Smolin referring to their concept of quantum gravity: They start with loops from the outset and show that since spin network states $<S|$ span the loop state space, it follows that any ket state $|\Psi>$ is uniquely determined by the values of the S-functionals on it, namely of the form

$$\Psi(S) := <S|\Psi>.$$

To be more precise, Rovelli and Smolin take embedded spin networks rather than the usual spin networks, i.e. they take the latter plus an immersion of its graph into a 3-manifold. Considering then, the equivalence classes of embedded oriented spin networks under diffeomorphisms, it can be shown that they are to be identified by the knotting properties of the embedded graph forming the network and by its colouring (which is the labelling of its links with positive integers referring to spin numbers).[42]

When generalizing this concept even further, a network design may be introduced as a conceptual approach towards pre-geometry based on the elementary concept of *distinctions, as* Louis Kauffman has shown.[43] In particular, space-time can be visualized as being produced directly from the operator algebra of a distinction. If thinking of distinctions in terms of 1–0 (or yes-no) decisions, we have a direct link here to information theory, which has been discussed recently again with a view to holography.[44] Ashtekar and Krasnov have noted this already when deriving the celebrated Bekenstein-Hawking formula in applying loops to black holes. Referring to punctured horizons they can show that each set of punctures gives actually rise to a Hilbert space of (Chern-Simons) quantum states of the connection on the horizon. Be $P = \{j_{p1} \ldots j_{pn}\}$ the set of punctures, and H_p the respective Hilbert space. Then $\dim H_p \sim \Pi_{jp \in P}(2j_p + 1)$, and the entropy of the black hole is simply given by $\Sigma_{bh} = \ln S_p \dim H_p$. The edges of spin networks can be visualized then as flux lines carrying area. With each given configuration of flux lines, there is a finite-dimensional Hilbert space describing the quantum states associated

---

der Substanz-Metaphysik,' *System & Struktur* I/1, 1993, 39–55; and 'Twistors & Substance. On Some Metaphysical Aspects of Science,' in: C.W. Kilmister (ed.), *Alternatives* (Cambridge 1994), 131–141.

42 C. Rovelli, L. Smolin, 'Spin Networks and Quantum Gravity,' gr-qc 9505006 (4/5/95). – If $D(\alpha)$ is the representation of the loop a, then the notation of Penrose can be recovered by $P(\alpha) = (-1)^{n(\alpha)+1} D(\alpha)$, where n is the number of single loops. In knot theoretic language, Penrose's spinor identity takes on the modified form $>< + X + \check{\times} = 0$.

43 L. H. Kaufmann, *Knots and Physics* (Singapore 1993), 459sq.

44 P.A. Zizzi, 'Holography, Quantum Geometry, and Quantum Information Theory,' *Entropy* 2 (2000), 39–69.

with curvature excitations initiated by the punctures of the horizon. Because the states that dominate the counting (for S) correspond to punctures all of which have labels $j = ½$, each microstate can be thought of as providing a yes-no decision or an elementary 'bit' of information.[45] Hence, the reference to Wheeler's 'It from Bit'. (Paola Zizzi has tried to generalize this within the conception of inflationary cosmology, and terms this 'It from Qubit'.[46])

Note that according to the standard terminology, a loop in some space Σ, say, is a continuous map γ from the unit interval into Σ such that $\gamma(0) = \gamma(1)$. The set of all such maps will be denoted by $\Omega\Sigma$, the loop space of Σ. Given a loop element γ, and a space A' of connections, we can define a complex function on $A' \times \Omega\Sigma$, the socalled *Wilson loop* such that

$$T_A(\gamma) := (1/N) \operatorname{Tr}_R P \exp \int_\gamma A.$$

Here, the path-ordered exponential of the connection $A \in A'$, along the loop γ, is also known as the holonomy of A along γ. The holonomy measures the change undergone by an internal vector when parallel transported along γ. The trace is taken in the representation R of G (which is actually the Lie group of Yang-Mills theory), N being the dimensionality of this representation. The quantity measures therefore the curvature (or field strength) in a gauge-invariant way.[47]

Over a given loop γ, the expectation value $<T(\gamma)>$ turns out to be equal to a knot invariant (the 'Kauffman bracket') such that when applied to spin networks, the latter shows up as a deformation of Penrose's value $V(\Gamma)$. This is mainly due to the fact that

$$<T(\gamma)> = K^k(\gamma) = (1/Z) \int d\mu(A) \exp(\ldots) T(\gamma, A).$$

So, for any spin network Γ (replace γ by Γ), the old relation holds up to regularization. Hence, spin networks are deformed into quantum spin networks (which are essentially given by a family of deformations of the original networks of Penrose). There is also a simplicial aspect to this: Loop quantum gravity provides for a quantization of geometric entities such as area and volume. The main sequence of the spectrum of area e.g., shows up as $A = 8\pi\gamma\hbar G \, \Sigma_i (j_i(j_i + 1))^{1/2}$, where the j's are half-integers labelling the eigenvalues. (Compare this with the remarks on black holes above.) This quantization shows that the states of the spin network basis are eigenvalues of some area and volume operators. We can say that a spin

---

45 Ashtekar, Krasnov, 'Quantum Geometry and Black Holes,' 14.
46 Zizzi, 'Holography, Quantum Geometry, and Quantum Information Theory,' 13.
47 I am following here the terminology as given by Renate Loll, in 'Knots and Quantum Gravity,' ed. J. Baez (Oxford 1994), 6sq.

network carries quanta of area along its links, and quanta of volume at its nodes. A quantum space-time can be decomposed therefore, in a basis of states visualized as made up by quanta of volume which in turn are separated by quanta of area (at the intersections and on the links, respectively). Hence, we can visualize a spin network as sitting on the dual of a cellular de-composition of physical space.[48]

As far as the dynamics of spin networks is concerned, there is still another, more recent approach, which appears to be promising as to the further development of topological aspects of quantum gravity (referred to as TQFT). In setting out to develop this new ansatz[49], John Baez notes that there are basically only two new ideas involved in loop quantum gravity. One is the insistence on a background-free approach. The other is to base the theory on the aspect of parallel transport rather than on the metric. Spin networks are at the basis of this approach. But although originally, Penrose thought of them in terms of describing the geometry of space-time, they really turn out to describe the geometry of space much better. The idea of Baez is therefore, to supplement loop quantum gravity with an appropriate path-integral formalism. While in traditional quantum field theory, path integrals are calculated using Feynman diagrams, he would like to introduce two-dimensional analogues of the latter, called *spin foams*.[50] Basically, a spin foam is a two-dimensional complex built from vertices, edges, and polygonal faces, with the faces labelled by group representations, and the edges labelled by intertwining operators. If we take a generic slice of a spin foam, we get a spin network.

Hence, a spin foam is essentially taking one spin network into another, of the form F: $\Psi \to \Psi'$. Just as spin networks are designed to merge the concepts of quantum state and geometry of space, spin foams shall serve the merging of concepts of quantum history and geometry of space-time.[51] Very much like Feynman diagrams do, also spin foams can be used to evaluate information about the history of a transition of which the amplitude is being determined. Hence, if Ψ and Ψ' are spin networks with underlying graphs γ and γ', then any spin foam F: $\Psi \to \Psi'$ determines an operator from $L^2(A_\gamma/G_\gamma)$ to $L^2(A_{\gamma'}/G_{\gamma'})$ denoted by O such that

$$<\Phi', O\Phi> = <\Phi', \Psi'><\Psi, \Phi>$$

---

48 C. Rovelli, 'Strings, loops, and others:...,' 8.
49 This idea goes back to M. Atiyah, *The Geometry and Physics of Knots* (Cambridge 1990).
50 J.C. Baez, 'An Introduction to Spin Foam Models of BF Theory and Quantum Gravity,' Preprint: 20/5/99. Here: 1sq. par.
51 J.C. Baez, 'An Introduction to Spin Foam Models of BF Theory and Quantum Gravity,' 32–34.

for any states Φ, Φ'. The evolution operator Z(M) is a linear combination of these operators weighted by the amplitudes Z(O). Obviously, we can define a category with spin networks as objects and spin foams as morphisms.

This leads to a discrete version of a path integral. Hence, re-arrangement of spin numbers on the 'combinatorial level' is equivalent to an evolution of states in terms of Hilbert spaces in the 'quantum picture' and effectively changes the topology of space on the 'cobordism level'. This can be understood as a kind of *manifold morphogenesis* in time: Visualize the n-dimensional manifold M (with $\partial M = S \cup S'$ - disjointly) as M: $S \to S'$, that is as a process (or as time) passing from S to S'. This is the mentioned *cobordism*. Note that composition of cobordisms holds and is associative, but *not commutative*. The identity cobordism can be interpreted as describing a passage of time when topology stays constant. If there is no change of topology (due to the action of the identity cobordism), then there is no change of state, because we do not have any local degrees of freedom here. Visualized this way, TQFT might suggest that general relativity and quantum theory are not so different after all. In fact, *the concepts of space and state turn out to be two aspects of a unified whole, likewise space-time and process*. Note that 'time' shows up here not as a function, but as a manifold (although arrows are used). This is particularly interesting, because with a view to what Barbour tells us about the 'absence' of time, this means that the concept of time is intrinsically included here as a pragmatic ordering principle for localizing topology changes. This is similar to what Prigogine calls the 'age of a system', which is roughly a frequency of formations of new structures in a system making the latter more complex. Time as a convention then, would be an approximate 'average' over such ages. (I have commented on this in more detail at another place.[52]) Hence, time shows up as being associated to a kind of measuring device for local complexity gradients. So what we have in the end, is a rough (and simplified) outline of the foundations of emergence, in the sense that we can localize the fine structure of emergence (the re-arrangements of spin numbers in purely combinatorial terms being visualized as a motion-in-itself) and its results on the 'macroscopic' scale (as a change of topology being visualized by physical observers as a motion-for-itself). This is actually what we would expect of a proper theory of emergence. But note also that space and time, in the classical sense, are obviously absent on a fundamental level of the theory, but can be recovered as concepts when tracing the way 'upward' to macroscopic structures. In other words: even as a gross average feature for 'shortsighted' human scientists (as Penrose indicates it at the end of his first twistor paper), space and time would nevertheless turn up as (philosophical) categories of concepts, simply, because the meaning of these categories is well-adapted to what humans actually perceive of their world (and communicate to

---

52 R. E. Zimmermann, *Selbstreferenz und poetische Praxis* (Cuxhaven 1991).

other humans). This is in fact, a point, where Barbour's argument seems to break down (if discussed within this philosophical perspective): What he essentially shows in his book is that quantum theory, *in so far as it is foundational,* describes partly what was called non-being (or substance) in former times. Hence, there is neither space nor time in *real* terms (= realiter, i.e. with respect to what there is in an absolute sense of the world's foundation), but there is space and time in *modal* terms (= modaliter, i.e. with respect to what humans perceive of their world). The former refers to substance, or, alternatively, to what *speculative* philosophy is all about. The latter refers to the physical world, or, to what *sceptical* philosophy is all about. The one relies on theoretical speculation according to what we know – speculating about the foundation of the world, which is outside (logically 'before') the world, and of which we are not a part therefore, and hence, about which we cannot actually know anything. The other refers to the empirical world, about which, with the help of experiments, we can obtain knowledge, in fact. Obviously, in terms of physics, the first (speculative) aspect is corresponding to physical theory, in so far as it is foundational. The second (sceptical) aspect corresponds to physical theory, in so far as it is empirical.[53]

These results can also be formulated in the language of category theory: As TQFT maps each manifold S representing space to a Hilbert space Z(S) and each cobordism M: S → S' representing space-time to an operator Z(M): Z(S) → Z(S') such that composition and identities are preserved, this means that TQFT is a functor Z: nCob → Hilb. Note that the non-commutativity of operators in quantum theory corresponds to non-commutativity of composing cobordisms, and adjoint operation in quantum theory turning an operator A: H → H' into A*: H' → H corresponds to the operation of reversing the roles of past and future in a cobordism M: S → S' obtaining M*: S' → S.[54]

---

[53] Do not think that both things would be the same for physics: In general relativity, we can clearly recognize that there is a well-defined conceptual part (in other words: a foundational part) which is actually clarifying the attitude of the approach, but without giving a pathway towards empirical results. But when the Einstein field equations are being introduced, then testable hypotheses can be formulated and checked in observational experiments. Probably, the same is true for quantum theory: In so far it is foundational, one would not expect any directions for experiments. Hence, Barbour would not have to defend his approach against Fay Dowker's objection of not having to offer any testable prediction. A fundamental theory is not actually obliged to make such an offer. Essentially, Barbour is telling us that there is no time at a fundamental level. But then, this is not a revolution at all, because in physics we learnt this already within the development of our century. In philosophy, already Spinoza has formulated this quite clearly, a long time ago.

[54] For a somewhat different approach to spin network evolution see also F. Markopoulou, L. Smolin, 'Causal Evolution of Spin Networks,' *Nucl.Phys.* B 508, 1997, 409, and id., 'Quantum Geometry with Intrinsic Local Causality,' gr-qc 9712067 (16/12/77). See also more recently F. Markopoulou, 'Dual Formulation of Spin Network Evolution,' gr-qc 9704013, id., 'The Internal Logic of Causal Sets: What the Universe Looks Like from the Inside,' gr-qc 9811053, and also id., 'Quantum Causal Histories,' hep-th 9904009 v4 (4/6/99).

4. Conclusion

So what do we have achieved after all? Spin networks, in particular as being visualized in terms of their quantum deformations, turn out in these models as the fundamental fabric of the world, in the sense that they are underlying and eventually actually producing the classical world, i.e. the world of classical physics as we know it from our everyday experience. On the other hand, these networks can be thought of as representing the most fundamental channels of information transport: It is in fact *quantum computation* which is permanently being performed through the channels the spin networks provide. The latter serve as a kind of universal lattice through which the information produced by quantum computation is percolating such that an eventual clustering in the sense of percolation theory spontaneously creates the onset of classicity. In other words, the computation taking permanently place within the fundamental spin network, lets the classical world emerge as we know it (with humans being a later – though transitory – product of this process themselves). The actual route taken is that via the formation of knots in terms of Wilson loops. That is, in the end, the classical world can be visualized equivalently as a 'condensation' of knotted spin networks.[55]

More recently, a similar path (but with different conclusions) has been taken by Stuart Kauffman and Lee Smolin: They basically utilize a deterministic model of directed percolation to achieve similar results and show that this can be visualized as a cellular automaton.[56] The idea is then to give the necessary criteria for a percolation phase transition which renders the system behaviour critical, as a derivation from Kauffman's idea of a 'fourth law' of thermodynamics. This idea is a fundamental one for Kauffman which he has elaborated upon in his large text on 'Investigations'[57]: He basically asks whether we can find a sense in which a non-ergodic Universe expands its total dimensionality, or 'total work space' in a sustainable way as fast as it can. He then refers to Smolin's interpretation of

---

55 The percolation problem is discussed in more detail elsewhere. For forthcoming work cf. my 'Classicity from Entangled Ensemble States of Knotted Spin Networks' (icmp 2000, Imperial College London, July, 2000) and 'Emergent Classicity: The Unfolding of Spin Networks: A Conceptual Approach to Self-Organized Criticality,' (MPI Dresden, May, 2000). The above mentioned results are also compatible with the concept of spin foams as put forward by John Baez. This will be shown in other forthcoming work by Paola Zizzi and myself: R.E. Zimmermann, P.A. Zizzi, 'The Functor Past Revisited: A Quantum Computational Model of Emergent Consciousness' and P.A. Zizzi, R.E. Zimmermann, 'Oncemore the Functor Past: Percolation Through Spin Networks and a Re-Interpretation of Spin Foams,' (2000) – Note that Louis Kauffman in his book on knots (quoted earlier) hints towards connections between Wilson loops and the topological information encoded in the spiral structure of the DNA.
56 S. Kauffman, L. Smolin, 'Combinatorial dynamics in quantum gravity,' hep-th 9809161 v2 (22/09/98).
57 S. Kauffman, 'Investigations,' http://www.santafe.edu/sfi/People/kauffman.

spin networks and their knotted structures at Planck scale level as 'comprising space itself'.[58] He suggests that knotted structures are combinatorial objects rather like molecules and symbol strings in grammar models, and he expects that such systems become 'collectively autocatalytic' – practically showing up as knots acting on knots to create knots in rich coupled cycles not unlike a metabolism:[59] 'The connecting concept will be that those pathways into the adjacent possible along which the adjacent possible grows fastest will simultaneously be the most complex and most readily lead to quantum decoherence, and classicity. If complexity "breeds" classicity, then the Universe may follow a path that maximizes complexity.'[60] This is in fact his concept of a 'fourth law': that the adjacent possible will be attained which could account for the explicit semantics and dynamics of evolution. Note however that this concept fits better to a fundamental structure which is non-local and a-causal rather than causal, and which we have discussed here in some detail. That is, classicity evolves from complete disorder represented by the fact that quantum computation deals with entangled qubits rather than with classical bits. Hence, it is very likely that classicity emerges from entangled ensemble states of knotted spin networks rather than from some structure which comprises of causal sets. But this is exactly what Kauffman and Smolin postulate as a consequence of their discussion: They visualize a history of a directed percolation problem as a causal set and conclude that it would have a causal structure 'and all its acutraments including discrete spacelike surfaces, light cones (!), future causal domains, past causal domains etc.'[61] Obviously, after the aforementioned, this is a rich causal structure on the fundamental level which turns out to be very unlikely after all. Note however that this refutation is mainly based on a conceptual argument rather than on a technical argument. And this is exactly the innovative aspect of modern philosophy providing for a critical reflexion of scientific results based on its heuristic potential. Hence, nowadays, the central task of philosophy is to conceptualize scientific results. Contrary to what the sciences do, namely to deal with their specific region of scientific knowledge their main activity is assigned to, philosophy has therefore the task to permanently

---

58 Kauffman, 'Investigations,' Lecture 7, 2.
59 Kauffman, 'Investigations,' par.
60 Kauffman, 'Investigations.'
61 S. Kauffman, L. Smolin, 'Combinatorial Dynamics in Quantum Gravity,' 12. – This refers to a strategy of conserving causality on a fundamental level of physics (despite the absence of space and time!) as being strictly followed in F. Markopoulou, L. Smolin, 'Causal Evolution of Spin Networks,' gr-qc 9702025, and a series of similar papers by both authors afterwards. See also F. Markopoulou, 'An Insider's Guide to Quantum Causal Histories,' hep-th 9912137. Compare however with L. Crane, 'Clock and Category: Is Quantum Gravity Algebraic?' gr-qc 9504038. Roger Penrose, in his Afterword (ch.32) of his birthday volume (S. A. Huggett et al (eds.), *The Geometric Universe*) formulates dryly that a 'final theory that emerges must have a fundamentally non-local character.' (424) If so, it must also be a-temporal and thus non-causal.

'totalize' all these regional results in order to achieve a consistent though transitory foundation for human orientation within the world. In this sense it is critical, and this is exactly what provides the close connection to an ethical foundation for human life.

Dedication

This work is dedicated to the memory of Alexandros Chapsiadis.

Acknowledgements

For illuminating discussions while in Cambridge I would like to thank John Baez, Julian Barbour, Richard Bell, Mary Hesse, Chris Isham, Lee Smolin, John Spudich, and Steve Vickers. In particular, I thank Paola Zizzi for important discussions and advice, and for her very kind and helpful co-operation in problems related to the topic presented here. I also thank Basil Hiley for his interest in this topic and the discussions we had about its implications, as well as for giving me the occasion to present parts of this conception to his physics theory group at Birkbeck College London.

13. Spinoza and Modern Physics

Comments on Professor Zimmermann's 'Spinoza in Context'

Juhani Pietarinen (Turku)

In his very rich paper, Professor Zimmermann considers Spinoza's metaphysics as a philosophical framework for recent attempts in theoretical physics to develop a unified foundational theory (Theory of Everything). The basic idea in such attempts is to specify a mathematical structure from which the description of the world as a material space-time system (the 'empirical' world) can be derived. This kind of foundational theory so to speak tells us the basic story of the physical reality. One candidate for a foundational theory is quantum gravity, according to which the physical reality ultimately consists of spin networks that produce all physical phenomena through permanent quantum computation.

Professor Zimmermann suggests that this kind of spin network with quantum computation might be taken as a physical interpretation of Spinoza's conception of substance (*natura naturans*) that produces all extended modal phenomena (*natura naturata*). Correspondingly, the theory of quantum gravity (if true) would represent adequate knowledge of this substance. He also suggests that this kind of foundational theory provides us with a ground for orientation in the world and thereby a basis for ethical life.

I will raise some questions concerning these suggestions.

1. For Spinoza, substance is something that is in itself and is conceived through itself.[1] That substance is 'in itself' means that its existence is necessarily caused by its essence; it has an infinite power to exist.[2] Does the spin network have this kind of infinite power of 'self-causation' so that nothing else is required for its existence? In other words: Is the theory of quantum gravity self-explanatory? In order to be so, it should state laws that are absolutely necessary and in need of no other ground than they themselves are. It seems implausible to claim that this would be the case, so that the quantum gravity does not really accord with Spinoza's necessitarianism, to which Professor Zimmermann refers (p. 167).

---

1   Spinoza, B., *The Ethics* (Indianapolis 1992), E1d3. E1 refers here to Spinoza's Ethics, part 1, and d1 refers to definition 1. Similarly, p1 refers to proposition 1, and p1s to scholium of proposition 1.
2   E1p34.

One consequence of Spinoza's necessitarianism (with his substance monism) is that this world of finite things is the only one that can follow from the nature of the unique substance. It is clear, however, that we cannot derive from the theory of quantum gravity that this world is the only possible one, because the abstract mathematical structure admits of different interpretations in terms of observable things. But if it cannot explain why just this actual world has been realized from the innumerable possibilities, it cannot be called, strictly speaking, a theory of *everything*. Professor Zimmermann describes what he calls 'quantum gravity substance' as having 'freedom to eventually produce a structured world'. In Spinoza's substance, however, there cannot be anything potential. The analogy to Spinoza's idea of substance seems to collapse here.

2. Spinoza's substance has the attributes of extension and thought.[3] Following Jonathan Bennett's suggestion[4], Professor Zimmermann takes the attribute of extension to be space and finite bodies as regional states or modifications of space (p 170). One point of Professor Zimmermann is that, in view of quantum gravity, temporality does not belong to the basic story of the physical world. 'Note that space and time, in the classical sense, are obviously absent on a fundamental level of the theory, but can be recovered as concepts when tracing the way "upward" to macroscopic structures' (p 182). We cannot, he says, speak of space and time in real terms but only in modal terms (p 182).

I find Zimmermann's point about space confusing. He seems to approve of the Spinozistic field metaphysics according to which the basic extended world (the attribute of extension) is *nothing but* space; but, on the other hand, he lets us know that in the foundational theory the notion of space is absent. The question arises whether space does belong to the *basic story* or not. Or do we have two different notions of space there?

What about the reality of time? In the foundational theory, Professor Zimmermann states, 'the concept of time is intrinsically included as a pragmatic ordering principle for localizing typology changes' (p 182). He seems to say that on the basic level there is no time but only an ordering of typological changes; time belongs to the empirical world alone. Professor Zimmermann takes this to accord with Spinoza's view, because, he says, time (*tempus*) is for Spinoza merely imagination and that 'temporality actually emerges in the transition from the infinity to finite modes (of thinking)' (p 170).

I would like to make two points here. First is an ambiguity in the notion of temporality in Spinoza. He does not say that the basic level of reality would lack all temporality. Because the substance, as well as its attributes, is eternal, it is last-

---

3   E2p2.
4   J.A. Bennett, *Study of Spinoza's Ethics* (Cambridge 1984), §§ 24–25.

ing through all times. This involves temporality in the sense of duration (lasting through time indefinitely).[5] This kind of temporality must be assumed if we think that changes and motions are real, as Spinoza thought. Where is no temporality, there is no change.[6] Thus, if a foundational theory of physics assumes an ordering of typological changes, it must also assume some kind of temporality. But this assumption does not imply any particular *measure* of time. It is true that for Spinoza all measures of time are arbitrary and 'imaginative'. If a foundational theory of physics assumes an ordering of typological changes, however, as Professor Zimmermann says, some kind of fixing of time must be assumed to make sense of temporal comparisons such as 'this region of space is now different from what it was before'. The question is then: In what sense is time absent on the basic level of physical world? Apparently the point is that no *definite* time appears there; the idea of dividing or cutting time is not involved. But, even on the most basic level we speak of changes, and therefore a concept of time in Spinoza's sense of *tempus* is needed.

My second point is concerned with Professor Zimmermann's assertion that in Spinoza temporality emerges 'in the transition from the infinity to finite modes' (p 170). This statement is in need of qualification. For Spinoza, actually existing finite modes have a definite duration (from their birth to their death), and in this they are unlike infinite things. All finite things, however, follow from the eternal substance[7], and, therefore, finite things also must have some kind of eternal (everlasting) existence, because whatever follows from the eternal essence of substance must be eternal.[8] Spinoza speaks of formal essences of individual things as contained in the attributes of the substance.[9] In this sense finite things have no definite duration, so that their (formal) existence does not involve any new kind of temporality. The actual existence of finite modes, however, involves a definite duration, and they are temporal in this new sense.

3. Another attribute of Spinoza's substance is thinking[10], a kind of power or ability of having 'ideas'. The same reality that is conceived as physically organized can also be conceived as mentally organized. How should we understand this?

Professor Zimmermann takes the position that for Spinoza only the extended ('material') world is real. He says that in Spinoza's view nature (as substance) does not think; only we human individuals think (p 167). This I think is mistaken.

---

5   See E2d5.
6   This point is made by Bennett, *Study of Spinoza's Ethics*, § 49.
7   E1p16.
8   E1p24.
9   In E2p8.
10 E2p1.

Spinoza says definitely: 'Thought is an attribute of God'[11]; '... thinking substance and extended substance are one and the same substance'.[12] This implies an overall parallelism that whatever is extended also thinks, and conversely. Every attribute expresses the essence of substance completely and independently of the other attributes; extension has no priority over thinking. This means that Spinoza does not endorse materialism.

Let us confine this reasoning to human individuals. Professor Zimmermann remarks that for Spinoza human nature is a special case of nature in general, so that 'every human action must be conceived as a manifestation of nature' (p 167). If nature is understood extensionally, only physical explanations of human thinking and acting are possible. This seems to be how Professor Zimmermann interprets Spinoza's view.

Spinoza explicitly denies, however, that any ideas (modes of thinking) can be caused by bodily states.[13] There is no causal influence between the attributes, so an explanation of ideas in terms of physical causes is impossible.[14] This is so despite that, as Spinoza says, 'a mode of extension and the idea of that mode are one and the same thing'.[15] Although mind is identical with body, the body does not cause any ideas, nor ideas any bodily affections. This view of Spinoza implies that mental phenomena are as natural as anything else in nature, but they can neither be reduced to or explained by physical phenomena.

Appealing to Tosel, Professor Zimmermann says that 'for Spinoza, all of reality emerges from another reality which is basically material' (p 169). If it is meant that mental reality emerges from the material one, this cannot be taken to be Spinoza's view.

One important consideration concerning materialism in general is the following. All materialistic or physical explanations of mental phenomena encounter the problem of representation. Our ideas (perceptions, beliefs, desires etc.) have a representational function; they have content. When I think of my friend, my thought somehow represents my friend, and to understand the thought, the content of my thought should, of course, be taken to be relevant. How can physical causal theories take such representative features into account? If they cannot, as seems to be the case, then they are not adequate for explaining mental phenomena and intentional human conduct.

---

11 E2p1.
12 E2p7s.
13 E2p6.
14 Michael Della Rocca investigates carefully the causal and explanatory barrier between thought and extension in Spinoza's system; see Della Rocca, *Representation and the Mind-Body Problem in Spinoza* (Oxford 1996), chapter 1, section III.
15 E2p7s.

It is interesting to note that Bennett attributes 'conjecturally' to Spinoza the view that (i) representative features of events have no part in physical causal theory, because pysical events are to be explained purely in terms of the shapes, sizes, positions, velocities etc. of the consistuent parts of the bodies, without referring to any representative features, and (ii) for any physical causal chain there exists an exact mental counterpart because of Spinoza's parallelism.[16] Hence, Bennett conjectures, Spinoza must regard the representative features of mental events as irrelevant to their causal powers, and this makes genuine teleological and intentional explanations impossible. I think, however, that Bennett's conjecture is not acceptable. A good reason for rejecting it is this: Spinoza makes it clear that there cannot be nothing from which a causal effect does not follow.[17] This means that things cannot have any causally irrelevant properties, and therefore also representational features of things must have causal effects.

4. My last remarks are concerned with the ethical point of view. As Professor Zimmermann rightly says, true knowledge of nature is according to Spinoza the basis for good life. It relieves one from passions and leads to freedom and peace of mind. The more we have adequate knowledge of nature, that is: the more we understand the nature of the substance, the more we are able to live in accordance with reason and virtue (which in Spinoza amount to the same things).[18]

Thus, if the final Theory of Everything were able to give us adequate knowledge of basic reality, it would provide us with a firm basis for ethical life. Indeed, in a sense, this follows from Spinoza's theory. When we understand the basic nature of the physical world, our mind is occupied by active thinking (reason) and the joy arising from it. *In that respect and to that extent* our mind is free. But in other respects our mind may suffer from many passions due to external causes. To get free from negative passions we have to understand them, that is, obtain adequate knowledge of the processes underlying them.[19] But, it is not plausible to think that any physical Theory of Everything, however adequate, would explain us how our passions arise in the particular situations in which we live. How does my knowledge of the spin networks help me to understand the sadness caused in me by the coolness of my friend? I need some other kind of knowledge relevant to this particular situation. Here again the representative aspect of knowledge plays a crucial role.

The problem is not just practical. I referred in the beginning to the problem of deriving explanations for particular facts from knowledge concerning the sub-

---

16   Bennett, *Study of Spinoza's Ethics*, §51.
17   E1p36.
18   E4p24.
19   E5pp1–20.

stance. I think we are faced with the same problem here, at least if the foundational physical theory is taken to explain the nature of the substance. Spinoza refers to a special kind of knowledge, *scientia intuitiva*, but his suggestions remain very obscure and cannot be taken as a basis for the scientific explanation of phenomena. If explanations of particular facts could be derived from a foundational theory, then we should conclude that it does not explicate the nature of the substance in Spinoza's sense.

14. Cosmology in Science and Theology

Some Perennial Differences

Stanley L. Jaki (South Orange)

Even at a cursory look a theological discourse on the cosmos should seem very different from scientific cosmology. And if one takes the utterances of the Bible about the cosmos for theological cosmology, a shocking clash may be on hand.

In the Bible the world resembles a tent.[1] The floor of that tent is the earth, which floats on waters and has a canopy, the firmament, over it. Celestial bodies, such as the sun, the moon, and the stars, are mere decorations fixed on that canopy. Such a world picture is worlds removed from the view science gives about the universe. This was true long before cosmology became a genuine science with the publication, in 1917, of Einstein's fifth and concluding memoir on the general theory of relativity.

In saying 'long before' I meant 'very long' indeed. Even the first guesses of the Greeks of old about the universe as a sphere with the earth in its center, were far superior to that biblical view in which the world was a mere tent. By taking the sky for a spherical container the Ionians (Thales and others) explained why the waters of the ocean stayed together. In the biblical world view the waters of the ocean are kept together by God and not by some physical factor or force. While nothing could be done scientifically with the world as a tent, the Ionians' universe was eventually developed into that fine scientific feat, which is Ptolemy's astronomy. It is well to recall that *The Almagest* of Ptolemy gave as accurate predictions of the motion of planets as did Copernicus' *De revolutionibus.*

By the time of Copernicus, Christians had for more than a millennium lulled themselves into believing that the Ptolemaic geocentric universe was somehow the same as the biblical world. They were very wrong. The biblical universe, a far cry from heliocentrism, was not even geocentric. In the biblical view the earth was not in the center of the universe but was the universe itself, because the sky was taken merely for the earth's canopy.

Some keen minds suspected that the biblical universe was not really equivalent to the geocentric universe. One of those keen minds was Augustine of Hippo. In

---

1 See on this the illustrations in my *Genesis 1 through the Ages* (Royal Oak, Mich. 1998), 250, 251, and 272, and chapter 1, 'Some Plain Irritants,' in my *Bible and Science* (Front Royal, Va. 1996) 19–31.

his *De Genesi ad litteram*, a long commentary on the Book of Genesis, Augustine warned that whenever a statement in the Bible about the physical world conflicted with what human reason reliably established about it, the Bible should be reinterpreted accordingly, lest the credibility of the Scriptures should suffer.[2] But even Augustine did not comply fully with this very sound principle. In view of the Bible's emphatic assertions about the firmament, Augustine took the firmament for a physical reality. Worse, he felt he had found it in a layer of vapor which, according to astrology, the 'cool' body of Saturn produced in its orbit.[3] At that time astrology was not yet separate from astronomy. It is well to recall that even Kepler and Galileo earned their living by casting horoscopes for their princely patrons.

There were various reasons why Christians lulled themselves into believing that there were no essential differences between the biblical world view and the Ptolemaic universe. One reason was that it is always tempting not to differ from the prevailing standard view. The other reason was even more important. In the geocentric universe the earth was in the center. Somehow this seemed to support a principal assertion of biblical cosmology, which was essentially a theological cosmology. According to it man was the center of God's attention inasmuch as God created everything for man's sake.

This was not a scientific proposition but a theological one. It was, of course, possible to argue that man was in the center of God's attention even if man was not in the center of a physical or cosmological manifold. There was room for a theological cosmology as long as one admitted metaphysics, or even some plain differences in the meaning of the use of the same word, such as center. But since metaphysics means something that follows physics, then as now it was pleasant to think that there is a direct line from physics to metaphysics. The line in question consisted in taking the idea of a geometrical center for a metaphysical center.

Such a line is illusory because it implies a conceptual reductionism. Apart from this it is possible to go from physics to metaphysics as long as one recognizes that some metaphysical propositions are indispensable in all physical theories. Physics must assume that there *are* things, that they are in *orderly* connection and so forth. But it is not possible to go in the opposite direction, namely, to go from metaphysics toward what is strictly scientific in a physical theory, or to the purely quantitative correlations contained in it. There is an irreversibility here between two conceptual lines or directions, just as there is an irreversibility in purely physical processes.

The line from a metaphysical center does not lead to a physical center. Conversely, the line from a physical center to a center taken in a metaphysical sense

---

2   In Book I, ch. 9. For details, see my *Genesis 1 through the Ages*, 86–90.
3   See my *Genesis 1 through the Ages*, 89.

may become illusory when the physical center proves to be an illusion. The latter process is well known. First the Earth was removed from the center of the system of planets through the work of Copernicus. Then doubts began to arise about the sun's central position after Galileo turned his telescope toward the sky and found a very large number of stars, especially in the Milky Way. More than a century later it was found that the Milky Way owed its appearance to the confluence of the light of an exceedingly large number of stars confined within a space that had the shape of a disk. But as late as 1900 it could still be believed that the Milky Way was the largest and also the central of galaxies and, since the sun appeared to be in the center of the Milky Way, the Sun (and the Earth with it) was still somehow in the center of at least the visible universe.[4]

Then in the 1920s it was found that this cozy galacto-centric universe was an illusion. The Andromeda galaxy turned out to be as large as our Milky Way. About the same time it was also found that our solar system was not in the Milky Way's center but about forty thousand light years from it. Then Einstein's four-dimensional cosmology imposed the recognition that no galaxy could be in the center because there was no center in a four-dimensional space-time. Clearly, nothing could be retained of the view that cosmic centrality in a metaphysical sense had a connection with cosmic centrality in a spatial or physical sense.

The picture which science gives now about the universe is not only vast but also very coherent. The same laws of physics, the same types of particles are known to operate over the awesome extent of about 60 orders of magnitude. This figure derives from the fact that science can trace the same processes of physics back to about 15 billion years, which is about 10^{17} seconds, and that within the so called first second, one can trace interactions to a fraction of it which is 10^{-43} seconds.

Those who now gloat over the discomfiture of the biblical or tent-like universe, taken for the basis for Christian or theological cosmology, should pause. They should reflect on three points. One is essentially a point of logic and has already been intimated by the difference between two very different meanings of the word 'centrality.' The second is that neither the Ptolemaic, nor the Copernican, nor the Newtonian, nor the Einsteinian cosmology can contain a refutation of two basic propositions of theological cosmology. One is the absolute, ontological dependence of all things on a Creator. The second is that all was created for the sake of man. Only if one assumes that one can evaluate these propositions in a quantitative way would it be possible to construct against them a scientific proof, which, let this not be overlooked, derives from quantitative verification.

---

4   See ch. 7. 'The Myth of One Island,' in my book *The Milky Way: An Elusive Road for Science* (New York 1972).

The third point may come as a great shock to unreflective admirers of modern scientific cosmology. A chief merit of that cosmology is that it removed the gravitational paradox that plagued Newtonian cosmology.[5] It did so by postulating a finite total mass placed in a four dimensional manifold. Yet, although the resulting picture suggests an over-all totality, the latter cannot be taken for that strict totality which the cosmos or the Universe, writ large, ought to be.

Even today the science of cosmology cannot demonstrate the existence of its object, the cosmos. Strangely, but truly, among all branches of science, scientific cosmology has the sad privilege of not being able to demonstrate the existence of its subject matter. Yet this should seem the primary task in any branch of science. In all branches of science, except one, the proof is very easy. It consists in one's merely opening one's eyes and then seeing the object. There is no need to prove the existence of the ocean, the object of oceanography. It is enough to take a boat, or fly over the ocean. There is no need to prove the existence of the Earth, in order to justify the subjects called geography, geophysics, and geology. There is no need to prove the existence of the air, it is enough to breathe, in order to provide the foundation for atmospheric science. And the same can be said of all other branches of science, with one exception: the exception is cosmology. Scientific proof is inseparable from observation and measurements. But it is not possible to place an instrument outside the universe so that it may be observed and measured.

And this is true even in this age of orbiting telescopes. It will never be possible to put a telescope in orbit around the universe in such a way as to position that telescope beyond the universe. Such a telescope will always remain within the universe because it can travel only along one or another of the paths of motion that are generated by the mass of the universe, that include the mass of the telescope itself. In order to do so, cosmologists would have to place a telescope beyond the universe in order to measure it, but this is an impossibility. Any gravitational body can only move along curved and re-entrant paths determined by the total mass assumed by the cosmological model and therefore remains part of that model. A proof of the existence of the universe can only come from metaphysics. All this, together with the construction of such a proof, was part of a series of lectures of mine given six years ago at the University of Liverpool, that were published under the title, *Is there a Universe?*[6]

In those lectures I also dealt with the shocking indifference which scientific cosmologists show toward all such problems so closely connected with their field of investigation. They think time and again that by choosing a new word, they

---

5   For details, see my *The Paradox of Olbers' Paradox: A Case History of Scientific Thought* (New York 1969). A new, entirely reset and enlarged edition is in the process of being published by Real View Books (Royal Oak, Michigan).
6   Liverpool University Press 1993.

can avoid a basic philosophical problem even when in the process they have to disregard the elementary dictates of science and logic. A case in point is the recent introduction of the term 'multiverse' into cosmology. Those have to do this who accept everything in the so-called inflationary universe. By everything I mean also some basic assumptions of the Copenhagen interpretation of quantum mechanics.

Quantum mechanics is a splendid form of statistical physics. The Copenhagen interpretation is a shabby philosophy, a strange mixture of pragmatism and idealism, a mixture concocted by Niels Bohr in Copenhagen. It implies the *non sequitur* that an interaction that cannot be measured exactly, cannot take place exactly. The *non sequitur* consists in justifying an ontological defect on the basis of an operational defect. Once this *non sequitur* is accepted, it is possible to claim that quantum mechanics can create things out of nothing, literally out of nothing. Details about the genesis and full unfolding of that claim can be found in my book, *God and the Cosmologists*.[7]

So much about the basis of the claim inherent in the inflationary theory that an indefinitely large number of universes are generated all the time. This is the justification of the use of the term 'multiverse' which is steadily gaining in popularity.[8] It would be in vain to confront the protagonists of multiverse cosmology with the problem of how an interpretation of physics can produce things literally out of nothing. This is a philosophical, an ontological problem, but it is of no concern to most scientists.

Champions of multiverse cosmology should therefore be challenged on purely scientific grounds: any good scientific theory must have predictions that are, at least in principle, verifiable by observation. Einsteinian cosmology is good scientific theory because it has such predictions. It can predict, from the value of the average density of matter in space, the total amount of matter that can be held in that cosmological model. It can predict the minimum curvature which a permissible path of motion can have in that model.

But the multiverse cosmology cannot predict the number of universes which it generates. Nor does it explain why those universes should not be governed with different laws and different physical constants. But an unlimited diversity of such constants and forces should be on hand if statistical considerations are applied rigorously. But then how could so different universes interact with one another? If, however, they cannot or do not interact, can a scientific knowledge be on hand about them?

---

7   See chapter 6, 'Loaded Dice.'
8   One of the term's prominent champions is Sir Martin Rees, who gave the title 'Intimations of a Multiverse' to the last of the four lectures he delivered at Princeton University in the spring of 2000. See also his *Before the Beginning: Our Universe and Others* (Reading, Mass. 1997) 3–4.

Clearly the multiverse theory is the ultimate phase of a development in scientific cosmology, which means the loss of the cosmos or the universe itself. Or rather it leaves the universe as an incoherent wreck. Therefore science can raise no objection to that basic proposition in theological cosmology that there is a coherent *all* or totality and that such a totality owes its existence to the Creator.

This basic proposition is the message of Genesis 1 or the first chapter of the first book in the Bible. That chapter has played an enormously informative part in Western thought but also has been the source of endless misunderstanding. Therefore I have to discuss it briefly. Is that chapter really a story about the making, about the genesis of the world? Is it really a cosmogenesis? For if it is, it cannot cope with scientific objections. Visible light, the product of the first day, cannot come before the sun, which appears only on the fourth day, nor can plants (that appear on the third day) perform their photosynthesis without the sun. There is no firmament.

But if Genesis 1 is a cosmogenesis, why does it end with a seventh-day rest of God? Surely, the Hebrews of old, who at least indirectly insisted that God was a pure spirit, could not assume that God needed some rest. It may therefore be more reasonable to assume that the story of Genesis 1 is a mere parable about the importance of the sabbath rest as a religious precept. If one assumes this, all details of Genesis 1 fall into a coherent whole that science cannot touch at all, a point that should seem of primary importance for a theological cosmology.

What I am going to say now very briefly can be found in great detail in a book of mine *Genesis 1 through the Ages,* first published in 1990 and reissued in a second enlarged edition in 1998.[9] The first seven of the nine chapters of this book are a review of the hapless efforts of theologians, over the last two thousand years, to reconcile that chapter with the science of the day. The last two chapters, where I present my explanation of Genesis 1 as a parable about the sabbath rest, can be read in brief in a little booklet of mine, 'The Sabbath Rest of the Maker of all: A Clue to Genesis 1.'[10]

Suppose then that God is set up to do some work through six days. The work has to be the highest conceivable work, which is the making of *all*. Second, God has to appear to work intelligently and effortlessly. Now any intelligent worker begins by providing light. This is what God does in the first day. He uses no tools, except his effortless word of command: 'Let there be.'

That God does or makes *all* is conveyed in three steps. First by the general statement: He made 'the heaven and the earth,' which is the classic Hebrew idiom for the *all*. Then the same is stated when God is said to have made the main parts,

---

9 See note 2 above.
10 Royal Oak, Michigan 1999.

the firmament and the earth (2nd and 3rd days). Again the same idea of totality is conveyed when the main decorations or details of the main parts (4th and 5th days) are listed as having been made by God. In all this we find the application of the hallowed rhetorical method of stating the whole by listing its main parts.[11]

That God works intelligently transpires not only from his starting with making light, but also from his providing at the end a manager of the edifice (world tent) he had constructed. As God's manager, man has to be the object of God's special care: man is made in the image of God. As to man's special formation from the clay of the earth, it should be reinterpreted according to the date of paleontology, just as the data of physics and astronomy had to serve above as guidelines.

In Genesis 1 taken for a parable, metaphysical assertions, physical reality, and moral instructions come together. The metaphysical assertion is that God made *all*, and therefore the *all* is wholly dependent on him. The physical reality is the *all*, as asserted in terms of its visible parts, the firmament and the earth. As long as these are taken for carriers of the message about the *all*, there is no conflict between science and Genesis 1. The *all*, or the absolute cosmic totality, let it be repeated again, cannot be observed by science. It is irrelevant whether one expresses the *all* in terms of a primitive image, a tent, or in terms of a scientific image, such as congeries of galaxies, or fundamental particles, or world lines, curvature of space and the like. To see that any of these may convey the meaning of a true *all*, one has to use one's metaphysical eyes.

The objection may arise whether all this is a belated projection into Genesis 1, or something that at least in some details has been seen from times of old? This is a question imposed by the historical method. The question is in a sense about the difference between the formulation of a correct method of interpretation, such as stated by Augustine of Hippo, and its implementation or rather the lack of it.

Of course, it was one thing to state a sound precept. It was another to apply it fully and consistently. Now here, theologians are no worse off than any other intellectual, including scientists. When theologians glimpse a timeless truth, they remain children of their own times, and so do scientists. As a result, important perceptions may go unexploited. Both the history of science and the history of theology or biblical exegesis provide many telling examples. About Genesis 1 it was noted on occasion that it had the teaching of the sabbath for its main idea and message. Yet this important insight failed to be developed. Instead of serving as a parable about the sabbath rest, Genesis 1 functioned as a cosmogenesis, which it never was meant to be.

Now regardless of whether one accepts the foregoing interpretation of Genesis 1 or not, any honest reader of that chapter must admit at least one thing: in Genesis

---

11   The method, *totum per partes,* is to be carefully distinguished from *pars pro toto,* which conveys the whole in terms of a principal part of it.

1 God is absolutely superior to all he makes. Surely, one does not find the idea of creation out of nothing in Genesis 1. Still one finds there God's absolute superiority over things, over all things, which is the very idea that supports the idea of creation out of nothing.

The idea of the absolute independence of God over everything and the absolute dependence of everything on God are strictly philosophical, indeed metaphysical notions. The theologian busy with cosmology should take note of this and also of the fact that science contains nothing of those ideas. Ontological independence and dependence are not notions that can be measured or observed with the physical eyes. To see those notions one must have philosophical eyes, just as one must use philosophy in order to see the universe as a strict totality.

The Christian or theological cosmologist can and must do what a scientific cosmologist cannot do as long as he remains within the limits set by the scientific method. The theological cosmologist has for his first task to demonstrate the existence of a strict totality of things if he truly wants to construct a theological cosmology. Since I dealt with this task in my Liverpool lectures, there is no reason to go here into particulars. The task is, of course, incumbent only on those theological cosmologists who take seriously the first statement in all Christian credal formulas: 'I believe in God the Father Almighty, Maker of Heaven and Earth, of all things visible and invisible.' About those theological cosmologists who worship the environment or the Gaia, I can merely say that instead of theology, they should rather call their subject poetical geology.

I must say more of a great opportunity, which a theological cosmologist should find in modern scientific cosmology. Partly, because I dealt with those opportunities in my book, *God and the Cosmologists*,[12] here too I can be brief. To see this opportunity one has to recall the nebular hypothesis of the origin of the universe as it was articulated by Herbert Spencer during the closing decades of the last century. According to him the primeval state was homogeneity which, by strange twists and turns, developed inhomogeneities. Of course, this is illogical. This is why Herbert Spencer furtively introduced slight inhomogeneities into that primeval homogeneity.

Now quite different is the picture given by modern scientific cosmology about the early and indeed earliest stages of the universe. The picture is a vast set of stark inhomogeneities, all of which can be expressed in numbers, in constants. The principal thing to note about constants is that they are constantly peculiar as all inhomogeneities are.

Herein lies the tremendous opportunity of the theological cosmologists. He should seize on those inhomogeneities, peculiarities, stark specificities and ask

---

12   See especially, ch. 3, 'Nebulosity Dissipated.'

the decisive and basic question: why such and not something else? But in order to raise this question, first phrased in this form by Leibniz,[13] the theological cosmologist has to realize that this is the question that governs the use of any bit of knowledge. In inquiring about anything we ask the equivalent of the question: why such and not something else?

Now in ordinary everyday things, their suchness can be traced to the suchness of other things. Suchness denotes dependence, that very biblical notion which we find in Genesis 1. But when it comes to the Universe, to its suchness, there are no other universes to which to refer the matter. Beyond the Universe writ large there can be only its Maker, its Creator.

Theological cosmology is not something that runs counter to scientific cosmology. The two cover two very different aspects of reality and go to two very different distances. The distance which science can go cannot go even as far as the Universe itself. Even the best scientific cosmological model is not necessarily equivalent to the Universe. And in case it happened to be equivalent, science can never know this. For two reasons. One is experimental, the other theoretical. The experimental reason simply states that it is not possible to assume that science would never discover such new data that would require a complete overhaul of the apparently final cosmological model.

The theoretical consideration is connected with Gödel's incompleteness theorems. According to those theorems, no non-trivial set of arithmetic propositions can have its proof of consistency within the set itself. Now a scientific cosmological theory must be non-trivially arithmetical because it has to be highly mathematical. Therefore whatever perfections that theory may have, it will lack an internal proof of consistency. This means that there can be no such thing as a scientific cosmology which would be true *a priori*. In other words, scientific cosmology cannot be used as an argument against the contingency of the Universe.[14] As far as we know, the Universe is not necessarily what it is. Therefore it could have been something else. Therefore, logic imposes the inference that the Universe, this actual and most specific universe, is the result of a choice among an infinitely large number of possibilities.

In exploring modern scientific cosmology, the theological cosmologist can find some genuine pieces of gold while finding also that scientific cosmologists can turn those pieces into fool's gold. An example of this relates to the fact that at its very earliest phases the mix of fundamental particles had to be exceedingly specific so that ultimately stars might develop within which carbon atoms, the basis of organic evolution, could form. Scientific cosmologists quickly called

---

13 He did so in his book, *Les principes de la nature et de la grace* (1716).
14 I set forth first this use of Gödel's theorem in my book, *The Relevance of Physics* (Chicago 1969), 127–30, and then in greater detail in *God and the Cosmologists*, ch. 4. 'Gödel's Shadow.'

this the anthropic principle as if the formation of carbon atoms would assure the eventual formation of human beings.

Of course only a fool would assume that such an eventuality would be a foregone conclusion. Only a fool would assume that evolution would repeat the same steps on our Earth, let alone elsewhere, if it were allowed to start over again. But the real foolishness in all this lies elsewhere. Some scientific cosmologists went on to claim that the universe exists because scientific cosmologists think about it. Surely there has never been a more foolish phrase than the one: 'I think therefore the universe does exist.'[15] Why is it then that just by thinking about a hundred dollar bill one cannot produce it?

While neither the scientific cosmologist nor the theological cosmologist can do such things, the theological cosmologist can go much further, infinitely further than his scientific counterpart. But in doing so the theological cosmologist must remain on the level of philosophy. And he must be careful not to mix supernatural viewpoints into what has to be above all a strictly philosophical discourse. If he does that mixing, the theological cosmologist destroys his philosophical edifice and ends in dreamland. A case in point is the vast set of dreams which Teilhard de Chardin offered with endless references to science, about whose branch known as physics he knew much less than he is supposed to have known.

As is well known, Teilhard de Chardin tried to see purpose in the great evolutionary process which according to him was leading to a noosphere, in which minds, far superior to the actually known human minds, would evolve. And he saw a cosmic Christ as the crowning of this purposeful development.

Teilhard de Chardin never considered some elementary problems that stare in the face of any historian of evolution and of Darwinism in particular. Now, within Darwinism, taken either for a philosophy or for a special mechanism of evolution, there is no room for speaking about purpose. There is no room, because this notion is barred at the outset from Darwinian philosophy and systematically excluded from its mechanism of evolution. Logic alone would impose the admission that just because a notion is excluded from a set of considerations constituting a philosophy or a scientific method, it therefore cannot follow that the notion is invalid or that it cannot correspond to something in reality.

Those who claim the opposite are guilty of fallacy in reasoning. Most Darwinists have blissfully ignored this. Whitehead was certainly justified when in 1929 he applied to them a phrase that cannot be quoted often enough: 'Those who devote themselves to the purpose of proving that there is no purpose, constitute an interesting subject for study.'[16]

---

15 The summary which T. Ferris gave of *The Anthropic Cosmological Principle* by J. D. Barrow and F. J. Tipler, in The New York Times Book Review, Feb. 16, 1986, 20.

16 A.N. Whitehead, *The Function of Reason* (Princeton 1929), 12.

Whitehead's phrase also applies to the dicta of some scientific cosmologists. A case in point is a statement in Steve Weinberg's book, *The First Three Minutes*. According to the grand conclusion of that book, the more we learn about the universe, the more meaningless it appears. Now this could only mean that cosmology finds no purpose in the universe. In the second edition of his book, Weinberg voiced his regret for having made that claim. But he did not say why he regretted it. Just because his claim provoked a barrage of criticism? Or was he reluctant to admit that his claim amounted to a rank disregard of the limits set by the scientific method?

Herein lie the basic and perennial differences between scientific cosmology and theological cosmology. When some prominent physicists tried to find evidence of a soul in physics, Maxwell, who firmly believed in the existence of an immortal human soul, warned those physicists by paraphrasing Shakespeare: 'There are many things in heaven and earth, which, by the selection required for the application of our scientific methods, have been excluded from our philosophy,' that is, science.[17] Among them are such elementary propositions as the assertion of the existence of plain everyday things and that when one asserts this one does it freely and for a purpose.

Those assertions are the fundamental philosophical assertions in any human discourse. Those assertions the scientist must assume although his method is unable to prove them. Nor does the scientist have to elaborate on them in order to do good science. On the contrary, the theological cosmologist must pay much attention to them, because only by unfolding their meaning can he speak of the cosmos in a sense in which the scientific cosmologist can never speak. For the scientific cosmologist, if he is consistent, there can only be a supergalactology. But the theological cosmologist must prove the existence of the universe, if he wants to use it as a jumping board to a level where his discourse about the cosmos becomes truly a theological cosmology. There one need not worry about measurements either of the cosmos or of anything else, because the subject is truly immeasurable. And there, if the theologian is a Christian theologian, he can justify the very foundation of the Christian Creed that makes no sense unless fully truthful is its first proposition about God the Father Almighty, Maker of Heaven and Earth, that is, the All, or the Universe.

---

17 See D. Niven (ed.), *The Scientific Papers of J.C. Maxwell* (Cambridge 1890) vol. II, 759.

15. An Axiomatic Foundation of Quantum Mechanics

Paradoxes, Causality, and the Unification of Sciences

Juleon M. Schins (Amsterdam)

1. Introduction

Often argued is that metaphysical arguments or suppositions should not be used in the interpretation of quantum mechanics. One should take the quantum formalism as it is and respect its foundations as they were laid down in the second half of the twentieth century. As the adjective 'metaphysical' is used in a largely negative sense – meaning something halfway 'unscientific' and 'primitive'– the above statements sound very convincing indeed. They would be so to my thinking as well, had the relatively short history of quantum interpretation not been plagued by an ever increasing number of 'paradoxes' and had one of the most important suppositions inherent to most present-day interpretations of quantum mechanics not been silenced. This supposition refers to the exact relationship between the wavefunction and the single event. *It is philosophically absurd to apply quantum mechanics to experimental result whatsoever – by definition the collection of single events – if there exists no relation between the quantum formalism and those very single events.* A clear formulation of the relation between the quantum formalism and the single event may not be an easy task, but without it philosophical consistency is not achievable. Surprisingly many authors are not only hardly interested in this relation, but even declare any interest in it as ridiculous. For example, Peter Mittelstaedt writes: 'Since neither of these attempts to understand probability on the individual level gives rise to any observable prediction, in the current literature the individualistic interpretation of quantum mechanics is not considered as an alternative that should be taken seriously.'[1] It seems that Mittelstaedt confounds physics with philosophy: quantitative prediction is the task of quantum mechanics, not the task of the interpretation of quantum mechanics. The latter's task is to provide a philosophical framework wherein quantum mechanics can be understood consistently. Consequently, an interpretation of quantum mechanics should not be judged on its added predictive force – quantum mechan-

---

1 Mittelstaedt, *The Interpretation of Quantum Mechanics and the Measurement Process* (Cambridge 1998).

ics does quite fine already, without the help of any interpretations – but on its consistency, both internal and with all other facts of ordinary experience. This idea does not seem to seduce many thinkers of the Copenhagen tradition, if we may believe Mara Beller:

> When Bohr speculated about parallels between wave-particle duality in physics and the complementarity of reason and emotion, or complementarity between different cultures, he asserted that the comparisons were not just vague analogies; *they flowed necessarily from the very analysis of the logical use of our concepts*. Bohr and his supporters presented his dualistic philosophy of complementarity in physics not as one feasible way of interpreting the quantum formalism, but rather as the only logically possible way. This rhetoric of inevitability implied the logical impossibility of any alternative to the Copenhagen philosophy, thus concealing the fruitful interpretive freedom of the quantum mechanical formalism. In this way, the philosophy of complementarity, while certainly legitimate as one of the many possible interpretive options, was turned into a rigid ideology, misleading both scientists and educated nonscientists.[2]

Sheldon Goldstein puts it even more drastically:

> Many physicists pay lip service to the Copenhagen interpretation, and in particular to the notion that quantum mechanics is about observation or results of measurement. But hardly anybody truly believes this anymore — and it is hard for me to believe that anyone really ever did. It seems clear that quantum mechanics is fundamentally about atoms and electrons, quarks and strings, and not primarily about those particular macroscopic regularities associated with what we call measurements of the properties of these things.[3]

Whatever the merits of the Copenhagen philosophy, it is a fact that nowadays alternative interpretations abound. The proliferation of these interpretations – hidden-variable, consistent histories, many-world, modal, and so many more – indicates that we are still far from a satisfactory solution. This is the reason that I have tried to analyze what elements in the traditional formulation may be considered the source of paradoxes. In order to do so, it was necessary to propose an explicit formulation of the relation between wavefunction and single event, as I think it is always implicit in the traditional interpretation. Together with four well-established axioms of the traditional foundation of quantum mechanics, this axi-

---

2 M. Beller, 'The Sokal Hoax: At Whom are we Laughing?' *Physics Today* 51/9 (1998), 29–34.
3 S. Goldstein, 'Quantum Theory without Observers (I)' *Physics Today* 51/3 (1998), 42–46.

omatic system can easily be seen to be wanting. Subsequently, I will propose an alternative relation between wavefunction and single event, and I will attempt to prove that it yields a foundation of quantum mechanics which

- allows for the completeness of quantum mechanics, and hence does not require nature to behave deterministically;
- removes all inconsistencies plaguing the traditional interpretation of quantum mechanics;
- and requires less assumptions than the traditional foundation.

The foundation proposed here is called Aristotelian, because it is inspired from Aristotle's hylemorphism. In that model, Aristotle considers a twofold causality in nature: material (*hyle*, meaning 'matter') and formal (*morphé*, meaning 'form'). Both causes are needed for the constitution of any material event, while neither of them can constitute an event on its own. Throughout this paper, causality is understood in the context of hylemorphism – as a necessary ground of being – unless otherwise indicated. This conception of causality resembles the non-temporal or 'essential' cause as conceived by Aristotle and Avicenna,[4] but differs from it in quite a number of aspects.

The second part of this contribution describes a theoretical unification of the sciences based on this adapted hylemorphism summarizing arguments elaborated elsewhere.[5]

2. *Clearing Out Paradoxes*

2.1. *The Traditional Foundation*

The world-view of the traditional interpretation is based on two main insights:

- without an observer there can be no event, only a mathematical evolution of superposition;
- scientific knowledge refers to how mankind sees, judges, or knows the world and not to how the world is or how it evolves.

The first tenet is at the heart of the Copenhagen epistemology. It is important to note that, although the single event is by definition observer-dependent, in the

---
4   T. Kukkonen, 'Causality and Cosmology: The Arabic Debate', in this volume.
5   J.M. Schins, *Hoeveel geest kan de wetenschap verdragen?* (Kampen 2000).

traditional interpretation the observer may not be considered as physically divisible from the system that is being measured: they form a unity. The second insight paraphrases a basic tenet of Kant's epistemology: according to Immanuel Kant, lawfulness is an addition of the human mind to the *noumenon*, which is in itself unknowable, non-causal, and non-lawful. These two insights form the basic suppositions of the world-view common to nearly all versions of the traditional interpretation of quantum mechanics.

The traditional foundation can be axiomatized as follows:

T1 every sufficiently isolated system has associated either a pure wave-function or a mixed state operator;

T2 the wavefunction obeys the Schrödinger equation as long as the system is not being measured ('freely evolving' system);

T3 the wavefunction evolves according to the projection postulate (wavefunction collapse) at the moment of measurement;

T4 the wavefunction bears a one-to-one correspondence with the physical system; that is to say, every change in the wavefunction (apart from an overall phase and amplitude constant) reflects a corresponding change in reality and *vice versa;*

T5 every detector has a classical dynamic variable (the pointer position) which bears a one-to-one correspondence with the system's dynamic variable that is being measured.

Proponents of the traditional foundation have never undertaken the formulation of a correspondence between wavefunction and single event, which they might even consider to be 'metaphysical' (i.e., useless). As a consequence, they will not accept axiom T4. Whatever the merits of the here presented formulation, it is quite obvious that it immediately yields all the paradoxes that are generally considered as inherent to the traditional foundation of quantum mechanics. The very proponents of the traditional foundation might question the terms or the status of axiom T4, but they will probably not deny its content.

The traditional axioms T1–T5 can easily be seen to be incomplete. For example, a sixth axiom is needed to specify when the pure wavefunction describes the system correctly, and when a density matrix should be used. This might seem an easy task. Maybe it is not difficult to prescribe some general rules, but that is not the goal of an axiomatic foundation. The additional axiom should be logically and philosophically consistent with the other ones. That is to say, any reference to 'reality', to 'apparatuses' and the like, should be consistent with the world-view in which the axiomatic foundation is embedded, and with the axioms themselves. Bohr himself excelled in the use of formulations that were either obscure or plainly inconsistent with the presumed axiomatic context.[6] A seventh axiom should specify when the measurement and corresponding wavefunction collapse occur. An eighth axiom should indicate what exactly is the difference between a given collection of atoms – organized in the form of an apparatus – and the 'classical apparatus' endowed with 'classical pointers' of non-superposed nature. What a formidable task! No surprise that seldom an axiomatic formulation is presented for the traditional foundation.

A particularly awkward axiom is T1, due to the ambiguous words 'sufficiently isolated'. This wording opens the way to many inconsistencies. Is a measuring device sufficiently isolated from the computer attached to it? If so, one may consider the wavefunction collapse to occur either on the detector surface, or at the detector output, or on the computer screen, or in the human brain... everything goes! Obviously, many authors quit the traditional foundation and consider the wavefunction collapse as purely mathematical, with no physical counterpart at all. This is a good first step, though nothing more than that.

'Sufficiently isolated' is not apt to quantification, while the question whether a wavefunction may be associated to something is digital: either yes or no. The direct interdependence of a qualitative and a digital concept is a guarantee for troubles. A similar argument also applies to the often heard tenet that quantum

---

6   J.S. Bell, *Speakable and Unspeakable in Quantum Mechanics* (Cambridge 1987), 189: 'However, Bohr went further than pragmatism, and put forward a philosophy of what lies behind the recipes. Rather than being disturbed by the ambiguity in principle, by the shiftiness of the division between "quantum system" and "classical apparatus", he seemed to take satisfaction in it. He seemed to revel in the contradictions, for example between "wave" and "particle", that seem to appear in any attempt to go beyond the pragmatic level. Not to resolve these contradictions and ambiguities, but rather to reconcile us to them, he put forward a philosophy which he called "complementarity". He thought that "complementarity" was important not only for physics, but for the whole of human knowledge. The justly immense prestige of Bohr has led to the mention of complementarity in most text books of quantum theory. But usually only in a few lines. One is tempted to suspect that the authors do not understand the Bohr philosophy sufficiently to find it helpful. Einstein himself had great difficulty in reaching a sharp formulation of Bohr's meaning. What hope then for the rest of us? There is very little I can say about "complementarity". But I wish to say one thing. It seems to me that Bohr used this word with the reverse of its usual meaning. Consider, for example, the elephant. From the front she is head, trunk, and two legs. From the back she is bottom, tail, and two legs. From the sides she is otherwise, and from top and bottom different again. These various views are complementary in the usual sense of the word. They supplement one another, they are consistent with one another, and they are all entailed by the unifying concept "elephant". It is my impression that to suppose Bohr used the word "complementarity" in this ordinary way would have been regarded by him as missing his point and trivializing his thought. He seems to insist rather that we must use in our analysis elements which either *contradict* one another, which do not add up to, or derive from, a whole. By "complementarity" he meant, it seems to me, the reverse: contradictoriness. Bohr seemed to like aphorisms as: "the opposite of a deep truth is also a deep truth"; "truth and clarity are complementary". Perhaps he took a subtle satisfaction in the use of a familiar word with the reverse of its familiar meaning.'

mechanics does not apply to systems containing many particles. If this is true, at what number of particles does quantum mechanics stop to apply? Or is the boundary continuous, implying that a three-body system agrees less well with the predictions of quantum theory than does a two-body system?

2.2. The Aristotelian Foundation

The Aristotelian foundation is best understood within a realist framework, which has two main ingredients:

- the existence of the world or of parts of it is completely independent of observation; in the wavefunction of the universe, observers play exactly the same role as all non-observing matter does;
- scientific knowledge refers to how the world is or how it evolves and not to how mankind sees, judges, or knows the world.

These ingredients are the very negation of those founding the traditional world-view. The realist world-view is quite unnatural to many of the early quantum workers; even to those who used some of Aristotle's ideas, like Werner Heisenberg, who suggested that Aristotle's act and potency might be relevant to quantum-measurement theory.[7] Heisenberg's concept of potency, however, (a status of reality halfway between mere possibility and actuality) has little to do with Aristotle's notion of potency. For Aristotle, potency was not a status of reality or of a single existing object, but the *coprinciple* of an existing material object. This coprinciple exists as objectively as the object itself, although it is not a self-sustained entity: it can only exist in conjunction with the act, the other coprinciple, which, in the case of all material things but human beings, is not a self-sustained entity either.

The Aristotelian foundation is founded upon the following axioms:

A1 there exists only a single exact wavefunction, which is the wavefunction of all existing matter; experimentally useful wavefunctions are approximated projections, valid within a restricted spatio-temporal interval, of this unique wavefunction;

A2 the 'wavefunction of everything' evolves according to the Schrödinger equation, without ever collapsing;

A4 reality is constituted by a twofold causality: material (the Schrödinger equation) and formal (choice); that is to say, every change in the wavefunc-

---

[7] T. Kallio-Tamminen, 'The Copenhagen Interpretation of Quantum Mechanics and the Question of Causality,' in this volume.

tion (apart from an overall phase and amplitude constant) reflects a corresponding change in reality, but not vice versa.

Due to axiom A4, the traditional axioms T3 and T5 can be discarded. Consequently, the concept 'wavefunction of the universe' is no longer inconsistent – as opposed to the traditional formulation – and the axiom T1 can be reformulated unambiguously. The present axiomatic system is very naturally compatible with Max Born's statistical interpretation of the expectation values. Born considered the quantum-mechanical probabilities as referring to objectively existing, but subjectively unknown, properties of a system, while the traditional foundation considers the values of observables before measurement objectively undecided. Evidently, Born's statistical interpretation is not tenable in the context of the traditional foundation.

In the following I shall briefly comment on five traditional paradoxes.

2.3. The First Paradox: the Artificial Boundary between System and Detector

The traditional foundation heavily relies on a strict boundary between the classical and quantum realms. Detector outputs are 'pinned down' and belong to the classical realm. The quantum realm, characterized by vagueness, encompasses all past processes that have not been observed. The traditional interpretation is not univocal concerning the degree of reality of the quantum realm (Bohr considered it non-existing at all[8]), nor does it adequately specify what the criteria are for observation.[9] These issues will probably remain obscure as long as one does not correctly distinguish physics from philosophy. The weakness of the traditional foundation is that it tries to impose a philosophical boundary – two different kinds of matter, or two different status of reality – on what is manifestly physical: the separation of one collection of atoms (the system) from an other one (the apparatus).

---

8 N. Bohr, quoted by A. Petersen, in M. Jammer, *The Philosophy of Quantum Mechanics* (New York 1974): 'There is no quantum world. There is only an abstract quantum-mechanical description.'

9 J.S. Bell, 'Quantum Mechanics for Cosmologists,' in C. Isham, R. Penrose, and D. Sciama (eds.), *Quantum Gravity 2* (Oxford 1981), 611–637: 'It would seem that the theory is exclusively concerned with "results of measurement" and has nothing to say about anything else. When the "system" in question is the whole world, where is the "measurer" to be found? Inside, rather than outside, presumably. What exactly qualifies some subsystems to play this role? Was the world wave function waiting to jump for thousands of millions of years until a single-celled living creature appeared? Or did it have to wait a little longer for some more highly qualified measurer – with a Ph.D.?' See also H. Stapp, *Mind, matter, and quantum mechanics* (Berlin 1993), 26.

The Aristotelian interpretation proposes but a single realm, which is 'classical' in the sense of 'pinned down' though not deterministically: the formal cause is not subject to quantitative lawfulness (the laws applying to formal causality are discussed in the second part of this contribution). Since in the Aristotelian interpretation the wavefunction never collapses, there is no need for postulating a boundary between a classical and a quantum realm: the boundary paradox literally vanishes. In this context it is important to note that the Aristotelian interpretation is perfectly compatible with the idea that measurement influences the system under study. But it does so in the same fashion as it does in classical physics. There, it is impossible to measure the temperature of a classical bath without influencing it, although (i) the influence can be made as small as one wishes, and (ii) even in the case of considerable influence, the properties of the disturbed system can in principle be reconstructed. The same holds for quantum measurements: as is long known, quantum mechanics even allows for interaction-free measurements, for which there seems to exist no classical analogue.[10]

In the Aristotelian interpretation measurement is conceived as a purely epistemological extraction of knowledge from reality, i.e., with no consequences for the overall wavefunction (including the observer) involved. *The measurement process certainly influences the system's evolution, but it does so in exactly the same way as unmeasured reality interacts.* Consequently, it makes no difference for the physics of the process whether a computer collects the measured data or not. This fact of ordinary experience is evident for everyone, but not a single version of the traditional interpretation is able to offer an explanation for it.

2.4. *The Second Paradox: Schrödinger's Cat*

Schrödinger's cat is the most famous paradox of the traditional foundation. It follows directly from axioms T1 and T4, the linearity of the Schrödinger equation, and the experience that reality is 'pinned down'. Consider an isolated, unobserved cat: according to axiom T1 it has associated a wavefunction. Dead and live cats are both known to exist, so that their corresponding wavefunctions must be valid solutions of the Schrödinger equation. Any living cat yields a non-vanishing probability that it either lives or does not within a short while. Consequently, the wavefunction of a living cat must, for its future evolution, mix progressively with the wavefunction of a dead cat. This evolution of the wavefunction implies a corresponding change in reality, according to axiom T4: the cat should be dying progressively. Ordinary experience, however, is quite definitive in confirming the

---

10 M. Renninger, 'Messungen ohne Störung des Messobjekts,' *Z. Phys.* 185 (1960), 417–421. See also the contribution by Elizur and Vaidman (note 11 below).

existence of either live or dead cats, thereby excluding quantum vagueness in the form of progressive mixing of opposites.

One might argue whether medical science does not offer some room for vagueness; sometimes it is not so easy to distinguish between life and death, certainly not on short time scales. But there can be no doubt concerning the definiteness of a cat's existing or never having existed at all. Consider the wavefunction of a male and a female cat and a bowl of food. The probability is that after some months this wavefunction has evolved so as to contain the addition of a baby cat, and the probability is that it has not – both probabilities summing up to unity are equally probable. If wavefunction and reality are related to each other as a mathematical bijection, i.e., with a one-to-one correspondence, one is at odds with the empirical fact that all things either happen *to be or not to be* – a fact of which even theater writers seem to be aware. So one has to be consistent and choose to abandon either the Schrödinger equation itself or axiom T4. The Aristotelian interpretation abandons axiom T4; and, on so doing, it solves Schrödinger's cat inconsistency on the most fundamental level.

2.5. The third paradox: interaction-free measurements

Interaction-free experiments allow the measurement of the existence of an object without energy exchange, so that the object in question suffers no change of quantum state. These interaction-free experiments sharpen the paradoxical nature of measurement in the traditional interpretation, especially since the early claims that any observation principally entails the exchange of energy or momentum quanta. Elitzur and Vaidman, conscious of this problem, comment:

> This paradox can be avoided in a framework of the many-worlds interpretation (MWI) which, however, has paradoxical features of its own. In the MWI there is no collapse and all 'branches' of the photon's state are real. These three branches correspond to three different 'worlds'. In one world the photon is scattered by the object, and in two others it does not. *Since all worlds take place in the physical universe,* we cannot say that nothing has 'touched' the object.[11]

Apart from the question whether MWI makes any sense at all – just think of how to axiomatize the concepts 'world' and 'universe' so as to be compatible with the enigmatic proposition since all worlds take place in the physical universe... –

---

11 A. C. Elitzur and L. Vaidman, 'Quantum Mechanical Interaction-Free Measurements,' *Found. Phys.* 23 (1993), 987–997.

it is not difficult to see that the Aristotelian interpretation has no problem with interaction-free measurements at all: the concept 'measurement' simply does not figure into the axiomatic basis.

2.6. The Fourth Paradox: Wave-particle Duality According to Wheeler

The wave-particle duality can be understood in many different ways. Here I will comment only on Wheeler's interpretation of his delayed choice double slit experiment.[12] Between slits and interference screen of an ordinary double slit experiment, Wheeler conceives of a large lens imaging the slits on the screen (see Fig. 1). On inserting the lens, the two slits are reproduced on the screen, and on withdrawing it, interference patterns appear. According to Wheeler, the time arrow of causality is invalidated by this thought experiment, because the experimenter can decide, *after* transition of the photon through the slit(s), whether it did so as a particle or as a wave. Obviously, Wheeler associates the lens-imaged impact of a photon with the passage of the photon through a single slit, and with the photon's particle character; on the other hand, he associates the photon's impact on the interference screen with the passage of the photon through both holes at the time, and with the photon's wave character. John Bell was truly amazed by these implications, saying 'it is better not to think about it.'[13] In the context of the Aristotelian interpretation, Wheeler's exercise in anti-causality is just one more consequence of the inconsistency of the traditional axioms.

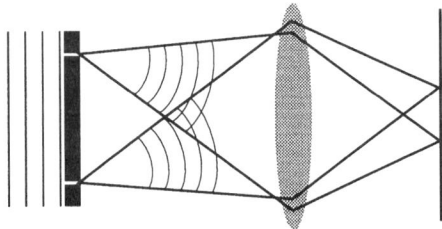

Figure 1: Schematic set-up of Wheeler's delayed choice double slit experiment. The lens is removable, such that the experimenter's decision to remove the lens can be taken after passage of the photon through the slit(s).

---

12 J.A. Wheeler, 'The Past and the Delayed-Choice Double-Slit Experiment,' in: A.R. Marlow (ed.), *Mathematical Foundations of Quantum Theory* (New York 1978), 9–48.
13 J.S. Bell, 'De Broglie-Bohm, Delayed Choice Double Slit Experiment, and Density Matrix,' *International Journal of Quantum Chemistry,* Quantum Chemistry Symposium 14 (1980), 155–159.

On replacing axiom T4 by A4, a photon is neither a particle nor a wave but an entity that has both aspects: always – i.e. independently from measurement and in a univocal fashion. These aspects are complementary, though not in the sense in which Bohr – who reveled in complementary incompatibilities[14] – used the word, but rather in its ordinary sense: such as the way an elephant's front and back are complementary. Axiom A4 states that a twofold causality is needed to constitute reality. The material part, given by the wavefunction, grants the photon its wave-like character, and the formal part, given by the choice, grants the photon its particle-like character. Both characters are always associated with the photon, whether measured or not. The wave-like character results in the photon's probability distribution. Inserting the lens may change the distribution in a quantitative way, but not philosophically: the disappearance of interference is a question of numbers, not of principle. The particle-like character assures that the photon always takes either one of the two slits and never both: this may safely be assumed, since a photon has never been measured to be at two different places simultaneously. Terribly enigmatic in the traditional foundation, and strikingly simple in the Aristotelian foundation.

2.7. The Fifth Paradox: The Arbitrariness of the Preferred Basis Set

Due to the linearity of the Schrödinger equation, the sum of two solutions is a solution as well. Together with the concept of the completeness of a basis, this idea implies that any solution of the Schrödinger equation can be written as a superposition of basis functions. Now consider the quantum mechanics of a complex composed of a system and measuring apparatus. From axiom T5 it follows that any initial state – pure or mixed – which is not a single basis function, will generally yield a final state which is a superposition of different pointer indications.[15] This is the source of the measurement problem in the traditional foundation of quantum mechanics. Alternative approaches, such as Zurek's,[16] may be promising for specific applications, but will not be of any significance in the solution of the preferred basis paradox as long as they are interpreted using inconsistent postulates.

In the context of the Aristotelian foundation, the preferred basis problem is solved by the consistency of the concept 'wavefunction of the universe'. The preferred basis follows directly from the way in which the many-particle wave-

---

14 See note 6 above.
15 B. d' Espagnat, *Veiled Reality: An Analysis of Present-Day Quantum Mechanical Concepts* (Reading, Mass. 1995), chapter 10; L.E. Ballentine, *Quantum Mechanics: A Modern Development* (Singapore 1999), chapter 9
16 W.H. Zurek, *Physics Today* 44/10 (1991), 36–44.

function of the universe is related to the single-particle wavefunction describing a given experiment. This is illustrated in the appendix for the case of a single particle experiment.

3. On the Unification of Sciences

3.1. The Crisis of Modernity

In the congress giving rise to this book, I presented the above axiomatic foundation of quantum mechanics as a response to Tarja Kallio-Tamminen's contribution on the Copenhagen interpretation of quantum mechanics. In an improvised introduction of the subsequent talk on causality, Jaakko Hintikka criticized the application of Aristotelian concepts for the foundation of quantum mechanics as obsolete, though without presenting any objections with respect to content. Three remarks are appropriate. First, since the foundations of quantum mechanics are a matter of philosophy and not of physics, one should not be surprised that twenty-four centuries old ideas might reappear. Secondly, the fact that my proposal is inspired on Aristotle does not imply the claim that Aristotle's biology or physics are still valid today. Thirdly, of Aristotle's hylemorphism I adopt only its very core: the twofold causality, related like act to potency. I define the material cause as the principle which is responsible for natural processes behaving in a quantitatively lawful way, and the formal cause as the principle which is responsible for the choices made in those natural processes. All associations of 'matter' presented by Aristotle, like it being the principle of individuation, and of total indetermination, lawlessness, and unknowability, of course do not apply to the here presented material cause. Quite the contrary: quantum mechanics clearly shows that it cannot account for individuation, and that it does account for a specific kind of indetermination (think of Heisenberg-like relations) though far from *total* indetermination. As far as lawfulness and knowability are concerned, evidently my material cause (the physical law) is the very opposite of what Aristotle had in mind in his hylemorphism.

Anyway, I admit that Hintikka's criticism, even when applied only to hylemorphism, agrees with a long tradition. As a matter of fact, the burial of Aristotle's hylemorphism began with René Descartes. The impressive scientific feats witnessed during the following three centuries seemed to confirm this burial, due to their deterministic character. Pierre-Simon de Laplace, the champion of determinism, proved that all the experimental data of celestial mechanics could be explained within a deterministic model, thereby implicitly declaring the twofold causality of hylemorphism superfluous. On the other side of the channel empiricism flirted with skepticism, thereby beginning the process that led inexorably to

an enormous discredit of philosophy in the eyes of scientists.[17] Today, scientists consider philosophy at best synonymous with (philo)logical text criticism, and the gap between the sciences and the humanities looms larger than ever.
Immanuel Kant is one of the last famous philosophers who seriously searched for a common foundation of science and the humanities. In view of this synthesis he postulated the existence of two levels of reality, the *noúmenon* and the *phenómenon*, the former being the ambit of morality and the latter that of natural sciences and causality. Kant's view on causality – the necessary relation between two events adjacent in time and space – was evidently inspired by the successful deterministic theories of his time, which he assumed to be definitive.[18] Kant's division of the world into two incommunicable parts implies a deep division in man himself. Not surprisingly, Kant finds it extremely difficult to account for the fact that a single human act can at the same time be moral and subject to deterministic laws.[19] The explanation that Kant finally proposes brushes verbal jugglery, rising

---

17   J. Laeuffer, 'Scientism and Scientific Knowledge of Things and God,' in: A. Driessen and A. Suarez (eds.), *Mathematical Undecidability, Quantum Nonlocality, and the Question of the Existence of God* (Dordrecht 1997), 185–192.
18   E. Schrödinger, *Mind and Matter* (Cambridge 1958): 'Kant's attitude towards science was incredibly naïve, as you will agree if you turn the leaves of his *Metafysische Anfangsgründe der Naturwissenschaft*. He accepted physical science in the form it had reached during his lifetime (1724–1804) as something more or less final and he busied himself to account for its statements philosophically. This happening to a great genius ought to be a warning to philosophers ever after. He would show plainly that space was necessarily infinite and believed firmly that it was in the nature of the human mind to endow it with the geometrical properties summarized by Euclid. In this Euclidean space a mollusc of matter moved, that is, changed its configuration as time went on. To Kant, as to any physicist of his period, space and time were two entirely different conceptions, so he had no qualms in calling the former the form of our external intuition, and time the form of our internal intuition (*Anschauung*). The recognition that Euclid's infinite space is not a necessary way of looking at the world of our experience and that space and time are better looked upon as one continuum of four dimensions seemed to shatter Kant's foundations – but actually did no harm to the more valuable part of his philosophy. This recognition was left to Einstein (and several others, H.A. Lorentz, Poincaré, Minkowski, for example). The mighty impact of their discoveries on philosophers, men-in-the-street, and ladies in the drawing-room is due to the fact that they brought it to the fore: even in the domain of our experience the spatio-temporal relations are much more intricate than Kant dreamt them to be, following in this all previous physicists, men-in-the-street and ladies in the drawing-room.'
19   I. Kant, *Kritik der praktischen Vernunft* (Frankfurt 1977), part VII (A170–171): 'Will man also einem Wesen, dessen Dasein in der Zeit bestimmt ist, Freiheit beilegen: so kann man es, so fern wenigstens, vom Gesetze der Naturnotwendigkeit aller Begebenheiten in seiner Existenz, mithin auch seiner Handlungen, nicht ausnehmen; denn das wäre so viel, als es dem blinden Ungefähr übergeben. Da dieses Gesetz aber unvermeidlich alle Kausalität der Dinge, so fern ihr Dasein in der Zeit bestimmbar ist, betrifft, so würde, wenn dieses die Art wäre, wonach man sich auch das Dasein dieser Dinge an sich selbst vorzustellen hätte, die Freiheit, als ein nichtiger und unmöglicher Begriff verworfen werden müssen. Folglich, wenn man sie noch retten will, so bleibt kein Weg übrig, als das Dasein eines Dinges, so fern es in der Zeit bestimmbar ist, folglich auch die Kausalität nach dem Gesetze der Naturnotwendigkeit, bloss der Erscheinung,

more questions than yielding answers. Since Kant, much progress has been made in the field of physics and mathematics. Not only have all the natural scientific ingredients of Kant's doctrine been shown to be flawed,[20] but, most importantly, we have witnessed for the first time in the history of science the formulation of a physical theory which is inherently non-deterministic: quantum mechanics. The completeness of quantum mechanics marks the end of Laplacian determinism and opens new avenues for approaching the unity of the sciences and the humanities. With inherently non-deterministic fundamental laws of physics no contradiction remains in a human act's being simultaneously moral and subject to physical laws.

3.2. Hylemorphism and the 'Postulate of Mind'

The application of hylemorphism to the human free act is straightforward: the material cause is provided by the Schrödinger equation, or whatever fundamental law applies to the processes under study. The formal cause fills in the choices left open by these fundamental theories, assumedly non-deterministic. As a consequence, the fact that one cannot swiftly travel to the Andromedae is due to the material cause, while the fact that one cannot kill an innocent person without being a murderer is due to the formal cause. These two causes are radically different though perfectly compatible; the material cause only determines the evolution of the wavefunction, while the formal cause chooses one of the many options allowed for by quantum mechanics. This solution fully respects the autonomy of both the sciences and the humanities without splitting reality into two incommunicative parts, as did Kant and Descartes in their own ways. It is probably the yawning chasm between Kant's and Aristotle's views on causality that – after

---

    die Freiheit aber eben demselben Wesen, als Dinge an sich selbst, beizulegen. So ist es allerdings unvermeidlich, wenn man beide einander widerwärtige Begriffe zugleich erhalten will; allein in der Anwendung, wenn man sie als in einer und derselben Handlung vereinigt, und also diese Vereinigung selbst erklären will, tun sich doch grossere Schwierigkeiten hervor, die eine solche Vereinigung untunlich zu machen scheinen.'

20 B. Russell, 'Logical Atomism,' in: A.J. Ayer (ed.), *Logical Positivism* (New York 1959), 31–52: 'In the first place, I found that many of the stock philosophical arguments about mathematics (derived in the main from Kant) had been rendered invalid by the progress of mathematics in the meanwhile. Non-Euclidean geometry had undermined the argument of the transcendental aesthetic. Weierstrass had shown that the differential and integral calculus do not require the conception of the infinitesimal, and that, therefore, all that had been said by philosophers on such subjects as the continuity of space and time and motion must be regarded as sheer error. Cantor freed the conception of infinite number from contradiction, and thus disposed of Kant's antinomies as well as many of Hegel's. Finally Frege showed in detail how arithmetic can be deduced from pure logic, without the need of any fresh ideas or axioms, thus disproving Kant's assertion that 7+5=12 is synthetic – at least in the obvious interpretation of that dictum.'

three centuries of deterministic triumphs – has sunk hylemorphism into deep oblivion among scientists.

In the following I will consider the consequences of hylemorphism and the associated realistic world-view for other scientific disciplines: mathematics, biology, ethics, theodicy, and theology. The study of these disciplines requires a second postulate besides hylemorphism. In my model, I consider the material and formal causes the only fundamental ones; unlike Aristotle, who conceived them on a par with efficient and final causality, and unlike Kant, who considered efficient causality as the only fundamental causality. This implies that true causality is, unlike the flawed billiard-ball causality of Kant and classical physics, transcendent. With other words, causality comes from outside the material universe, and is therefore in need of a source. This source must evidently be transcendent as well. Hence, it is quite in line with the here proposed variant of hylemorphism to postulate such a transcendent source of causality. In order to account for human freedom, I propose to multiply this source, as in the following *postulate of mind: with every human person is associated a source of formal causality which is the seat of individuality, freedom, (mathematical) understanding and creativity.* This source of formal causality does not provide all formal causality within man (most quantum choices in one's body are beyond one's free will) but it does influence those physical processes in the brain which are of fundamental importance for the expression of human freedom.

3.3. Mathematics

Roger Penrose has argued quite convincingly that a Turing computer is principally unable to simulate human thinking and, specifically, the intellectual grasp of Gödel's incompleteness theorem.[21] Why is this? Penrose searches the answer in a new physical theory encompassing both quantum mechanics and gravity. His speculations concerning coherence and wavefunction collapses in the brain have been severely criticized and recently disproved quantitatively.[22] Clearly, in the context of hylemorphism, no new physical theory will ever explain the superiority of human thinking over computers:

- the calculation of exact solutions performed by computers is founded on the deterministic aspect of the behavior of matter – fuzzy logic is of no help here, since Penrose is concerned with explaining why scientists necessarily agree on mathematical identities or lawfulness;

---

21 R. Penrose, *The Emperor's New Mind* (Oxford 1989), 105.
22 C. Seife, 'Cold Numbers Unmake the Quantum Mind,' *Science* 287 (2000, March 16th), 791.

- the deterministic aspect of the behavior of matter is precisely that aspect of reality that can be described by mathematical equations, the genuine expression of physical laws.

In other words: any new physical law, however revolutionary, will always be written in the language of mathematics and will consequently form the basis of possible computational operations. Penrose's argument is stated in very general terms and must, therefore, hold for all possible natural laws which are founded axiomatically. But no evidence allows that future laws of nature will cease being formulated as quantitative relations in the context of an axiomatic system. As a consequence, natural lawfulness (material causality in Aristotelian terminology) will never explain the fact that only human beings are able to grasp the truth of Gödel's incompleteness theorem.

The postulate of mind, in the context of Aristotelian hylemorphism, is an effective way out: mathematical understanding is not something that can be attributed to any physical object, regardless of how fast it may flop its many bits, but only to physical objects that have their own, non-material mind as a source of formal causality.

3.4. *Biology*

The twentieth century has made clear in what way organisms pass on, from generation to generation, all the information needed for their construction: by means of DNA molecules. From the software business is known that *useful* programs, differing slightly in source code, also differ slightly in performance. In other words, it is impossible to write two programs with a completely different application (for example, tax calculation versus optimization of transport) in such a way that the software codes differ in only 1 % of the coding bits. Analogously, biological organisms differing slightly in genetic code also differ slightly in performance, power of expression, or 'richness of behavior'. Many examples can be found in Dayhoff's *Atlas of protein structure and function*:[23] for the protein cytochrome-c, a comparison of organisms ranking from bacteria through humans is available. Different yeast species differ about 25 % in their cytochrome-c protein sequences, while closely related insects, fish, or mammals may differ less than 1%, unrelated species always differing substantially. Exactly the same occurs in software: subsequent releases of a given program bear great resemblance in performance and code, although analogous performance may be reached by totally different source codes, while totally different performance always implies totally different source codes.

---

23 M.O. Dayhoff, *Atlas of Protein Structure and Function* (Silver Spring MD 1972), volume 5.

The paradox in evolution theory arises from the two well-known facts

- that man and chimpanzee differ by only 1.35 % in their genetic content,[24] or, which amounts to exactly the same, that man and chimpanzee have a very recent common ancestor – about 5 to 8 million years old;[25]
- *that man is the only organism which is capable of unlimited intentionality and abstraction, while chimpanzees do not reach beyond second order, and species lower than mammals not beyond first order.*[26]

One may find it remarkable that nature needed one billion years to build a living cell and three billion more to build a fish. Stunning is the fact that it needed nearly all of evolution time available to build an ape – able of second order intentionality and abstraction – but a mere five million years more (approximately 1000 times shorter) to build man – capable of infinite order intentionality and of abstracting the very process of abstraction itself. It seems improbable that biology will, in the near or distant future, come up with an alternative information carrier, parallel to the DNA molecules. Consequently, biology will probably not be able to explain the absolute uniqueness of human behavior among all other organisms. Again, the postulate of mind offers a solution. Between all organisms and man is a fundamental discontinuity: the non-material mind, source of intelligence, consciousness, and free will. This mind explains why man, genetically quite close to the chimpanzee, differs so spectacularly in his behavior.

3.5. Ethics

The difference in ethical behavior between man and chimpanzee may at first glance seem only gradual, but it is not. Like all other animals, chimpanzees have their codes and patterns of behavior, chimpanzees being the most differentiated species.[27] Although many of these codes are as yet poorly understood, nothing indicates that their origin must be sought outside evolution. In other words, animal behavior can in principle be reduced to natural laws, formulated in disciplines such as physics, chemistry, biology, psychology, and sociology.

Human behavior seems to be unique among all mammals in that human acts have a direct repercussion on the actors themselves. For example, sexual abuse

---

24 L.S. Whitfield, J.E. Sulston, and P.N. Goodfellow 'Sequence Variation of the Human Y-Chromosome,' *Nature* 378 (1995), 379–380.
25 B. Wood and A. Brooks, *Nature* 400 (1999, July 15th), 219–220.
26 D. Premack and A. Premack, *The Mind of an Ape* (New York 1983).
27 A. Whiten, J. Goodall, W.C. McGrew, T. Nishida, V. Reynolds, Y. Sugiyama, C.E.G. Tutin, R.W. Wrangham and C. Boesch, *Nature* 399 (1999 June 17th), 682.

is a quite common phenomenon among mammals; but only among humans is it evident that the perpetrators of sexual abuse are social outcasts. This is not at all the case for chimpanzees. The same holds for murderers: in human society they are deeply frustrated people (even though they might appear superficially to be happy) while in chimpanzee society they show no sign of unhappiness. In the context of philosophical materialism human behavior should be reduceable to genes. But what is the evolutionary advantage of genes that make a murderer feel miserable, a sexual abuser being cast out of society, and why do those very genes work out so differently in chimpanzees?

Present day ethical theories (mostly consequentialist or deontological) are not at all concerned with the origins of human morality. Their proponents have not the slightest idea how to motivate an individual to act morally. Kant says one simply *has* to do so, but who cares what Kant says in the face of a strictly personal moral choice? The consequentialists say societies with altruistic behavior have higher survival probabilities; but again, who cares about society in the face of a strictly personal moral choice? The above-mentioned ethical theories are unable to explain the most trivial facts of human moral behavior, and must therefore be considered as theories with a largely applied character, still in need of a proper theoretical foundation.

In the context of Aristotelian hylemorphism and the postulate of mind the uniqueness of human behavior can be explained as follows: just as the material aspect of reality obeys mathematical laws, the formal aspect of reality obeys ethical laws. An example of an ethical law is the following: the murderer of an innocent person damages his spiritual health. Such damage could, for example, consist in the weakening of both free will and cognition of what harms or what profits (think of drug addiction). In much the same way that material laws, if general enough, *define* matter, ethical laws, if general enough, *define* the human mind.

3.6. Theodicy

Of great importance is that the cosmological arguments for the existence of God be rediscovered by present-day philosophers of religion.[28] In this contribution I have argued that a single kind of causality (call it material) does not suffice to explain fundamental problems in physics, biology, and mathematics. But even if there were only one kind of causality responsible for our universe, it still has to be explained scientifically. If everything is matter, then the fundamental laws of physics are also matter, which is absurd. The alternative – that fundamental

---

28   K. Yandell, *Philosophy of Religion: A Contemporary Introduction* (London 1999).

laws of physics do not exist – is just as absurd. Apparently, hylemorphism yields a solution in cosmology as well.

Once it is agreed that causality needs an explanation, the intellectual journey towards the necessary existence of an uncaused cause is not really a long one. What is more involved is how to understand the existence of moral evil in the face of a supposedly almighty God: the original problem of theodicy as studied by Gottfried Wilhelm von Leibniz in his *Die Theodizee von der Gute Gottes, der Freiheit des Menschen und dem Ursprung des Übels*. Alvin Plantinga solves the problem of theodicy by carefully analyzing the concepts of essence, possible world, and freedom.[29] Although the last word has not been written or spoken, I am not aware of any convincing argument against Plantinga. It is probably needless to add that Plantinga's argument is fully compatible with both hylemorphism and the postulate of mind.

3.7. Theology

There are as many definitions of theology as there are theologians. Although I am not one myself, I will not refrain from adding one more definition to the files. For this purpose, I will first consider in what way sciences and humanities can be unified. Thanks to hylemorphism, we have a straightforward explanation on one hand of the unity of reality and on the other hand of the radical difference between two kinds of disciplines: natural sciences focus on material causality and formulate quantitative laws; humanities focus on formal causality, and formulate qualitative laws. This sounds extremely simple – and it is indeed so – but it should not be forgotten that not a single philosophical model has ever been able to explain these two features: the unity of reality and the radical difference of the studied laws. However, a true unification of the sciences also needs a unique criterion to distinguish scientific knowledge from ordinary knowledge. In the framework of hylemorphism, this criterion is achieved as follows: scientific knowledge obtains when a set of complex phenomena can be reduced to a small set of axioms. Such a reduction defines the scientific explanation. As illustrated above, this criterion applies to natural sciences as well as to humanities.

One may ask whether theology also fits in this unified picture. It does so, in my view, on the condition that the scope of sciences be widened to contemplate *all facts,* instead of only empirical facts. In this case, theology would be a true science if (i) it considers *facts,* and (ii) it explains these facts by axiomatic reduction. Since theology does not consider empirical facts but the content of divine revelation, it is impossible to determine scientifically which are the facts with

---

[29] A. Plantinga, *The Nature of Necessity* (Oxford 1974).

which theology should be concerned. This problem does not imply, however, that theology is unscientific but only that it is quite a risky business. Consequently, theology appears, in this unified foundation, as a truly scientific discipline. It would not, however, be just one more science, like mathematics, biology, ethics or economy. Since it considers a set of facts (those from divine revelation) that is essentially out of the reach of the natural sciences and the humanities, theology is in a sense the complement of all other sciences put together.

Appendix: Relating the Wavefunction of the Universe to the Wavefunction of the Experimental Physicist

The relation between the wavefunction of the universe and that of the experimental physicist cannot, of course, be derived rigorously for any possible experiment. For this reason I will consider here an archetypal experiment: consider a single-orifice dark room containing a light detector, where the orifice is much smaller than the frequency of the light for which the detector is sensitive. In an approximate single particle description, the initial single particle wavefunction $|i>$ and single particle Hamiltonian H define the probability P that after a time interval δt the particle ends up in the single-particle state $|f>$ as $P=|<f|q>|^2$, with $|q>$ the single particle propagated wavefunction, given by $|q>=\exp(-iH\delta t)|i>$. The single particle wavefunctions $|i>$ and $|f>$ are both members of an orthonormal basis set $\{|b_k>\}$. This is the experimental physicist's description of reality.

The wavefunction of all matter $|\Psi>$ may be expressed in so far as the world's photons are concerned as a superposition of products of those basis functions:

$$|\Psi> = \Sigma_{k1,k2,...kN}\, c_{k1,k2,...kN}\, |b_{k1}>|b_{k2}>...|b_{kN}>,$$

with N representing the total number of particles in the universe. The index k1 in the product refers to particle 1, the particle under study. The probability Π_i that $|\Psi>$ has, at some initial time t_i, particle 1 in initial state $|i>$, is given by

$$\Pi_i = \Sigma_{k2,k3...kN}\, |<\Psi(t_i)|i>|b_{k2}>...|b_{kN}>|^2,$$

the first particle being excluded from the sum. Similarly, the probability Pf that Y has, at some time $t_f=t_i+\delta t$, particle 1 in state $|f>$ is given by

$$\Pi_f = \Sigma_{k2,k3...kN}\, |<\Psi(t_f)|f>|b_{k2}>...|b_{kN}>|^2,$$

the first particle being again excluded from the sum. The single particle probability P is an approximation of the exact N-particle expression $\Pi=\Pi_f/\Pi_i$. In general, this expression does not represent a conditional probability, but in our archetypal experiment, where a measured photon must necessarily have passed through the orifice, the mentioned ratio does represent a conditional probability.

We now write the wavefuction of the world at time t_i (when the photon is exactly in the orifice) as the sum of two terms

$$|\Psi(t_i)> = |\Psi_1(t_i)> + |\Psi_2(t_i)>$$

where

$$|\Psi_1(t_i)> = \delta(x_1-x_i)\, \delta(t-t_i)\, \Sigma_{k1,k2,..kN}\, c_{k1,k2...kN}(t_i)\, |b_{k1}>|b_{k2}>...|b_{kN}>, \text{ and}$$
$$|\Psi_2(t_i)> = (1-\delta(x_1-x_i)\, \delta(t-t_i))\, \Sigma_{k1,k2,..kN}\, c_{k1,k2...kN}(t_i)\, |b_{k1}>|b_{k2}>...|b_{kN}>.$$

The delta functions vanish everywhere but in the orifice at time t_i. Due to these delta functions, all x_1-dependent functions $|b_{k1}>$ may be discounted by introducing a new set of coefficients $d(t)$, thereby allowing for the elimination of k1 from the summation:

$$|\Psi_1(t_i)> = \delta(x_1-x_i)\, \delta(t-t_i)\, \Sigma_{k2...kN}\, d_{k2...kN}(t_i)\, |b_{k2}>...|b_{kN}>.$$

The delta functions now provide the initial state $|i>$ of the experiment; consequently, in the calculation of the probability Π_i, the term $|\Psi_2(t_i)>$ is immaterial. The wavefunctions $|\Psi_1>$ and $|\Psi_2>$ evolve in time according to the Hamiltonian of the universe H. This Hamiltonian may be written as a sum of three terms:

$$H = H_1 + H_{2-N} + V.$$

Here, H_1 contains only operators of the first photon, H_{2-N} only operators of photons 2 to N; V represents the interaction between the first photon and all the other photons. For photons this interaction happens to be zero, but for other particles, such as electrons, which carry a far-reaching Coulomb field, this is not necessarily the case. Had the interaction vanished, then the initial wavefunction would have evolved only under H_1 and the many particle problem would have yielded exactly the same probability as the single particle approach.

This situation is what I want to show to be the case very generally: also in the case of electrons. What then is responsible for the fact that, even in the case of electrons, the evolution of the measured electron is determined by H_1 alone and not by the interaction V? The reason lies in the careful preparation of the experiment. As a matter of fact, there is a second factor in the wavefunction of the universe, which until here described only the photons, that describes all other matter, part of which serves as the shielding (the walls of our archetypal experimental box).

When we include this matter in the wavefunction of the universe as well as the corresponding boundary conditions in the initial and final wavefunctions, the probability expressions Π_i and Π_f imply the projection of the evolved wavefunction of the universe, containing all possible superposed alternatives, on the deltafunctions of the initial and final wavefunctions characterizing the shielding: in our case, the location of the walls of the single-orifice dark room. The terms in the wavefunction of the universe are exactly those for which the above mentioned interaction V approximately vanishes: the better the experiment is performed, the smaller the interaction V. Therefore, for a well designed archetypal experiment, the single particle probability P is a correspondingly good approximation for the N-particle expression $\Pi=\Pi_f/\Pi_i$. This argument, of course, is not restricted to single particle experiments nor to experiments with just a single kind of particle. Nor is this argument meant to be rigorous, as is evident from the fact that I did not take the interaction between photons and matter into account.

16. Cosmology, Theology, and the Question of Ultimate Explanation

Heikki Kirjavainen (Helsinki)

It has sometimes been said that Oriental thought differs fundamentally from its Western counterpart because Oriental thought aims at unity while Western thought is more appreciative of analysis, ie. separateness. The claim is simplistic, but it is true that Western thought typically offers more structured studies of reality than Oriental modes of thought do. 'The great chain of being' in Western thought has been conceived as a series of attempts to precisely demarcate conceptually the dividing lines between man, world, and God. The division between man and the world (nature) has often been thought to rise out of cognitive and above all, scientific considerations. This cannot but have influenced the fact that religion, too, man's relation to God, has often been viewed in cognitive terms. Consequently, studies of the relationship between God and nature have not generally been considered as independent but as subordinate to the cognitive issues of religion and science.

In this paper I shall move between certain historical and contemporary themes in metaphysics and cosmology. All of these have to do with the dream of the final intelligibility of reality. I will try to shed light on how ancient presuppositions resurface in the contemporary discussion. Particularly interesting in this regard are notions that concern the allegedly isomorphic character of reason and true being as well as notions having to do with contingency and necessity.

1. *The Great Shifts in Metaphysics: from Participation to Interpretation*

The new way of understanding nature as contingent, i.e., as non-necessary, constituted a monumental conceptual shift in the late Middle Ages. Formerly, reference to those properties that made beings what they are had sufficed to explain them. Because these essential features were taken to be metaphysically exhaustive and intelligible, scholastics before John Duns Scotus and William Ockham did not put appreciable emphasis on contingent invariances (what we would call causal relations) either metaphysically or in broader outlook. Logical or metaphysical invariances were deemed adequate for a comprehensive understanding of the world. In this view, reality was understood looking backwards to the past. Here, being shows itself as frozen in invariance and necessity. All one has to do in order to gain an accurate intellectual view of reality is to differentiate beings from each other by means of their unchanging specific features. It was furthermore assumed that personal perspective does not alter the objects of knowledge.

If observation attends to the present, however, an understanding of the world's universal features seems to require new kinds of invariances: mathematical invariances of types of change, to be formulated on the basis of experience. The observer's own position and point of view also seems to become markedly significant. One could summarise the difference between the two attitudes as follows. In the earlier approach one proceeds from the metaphysical essence of the creature to establish what kind of a cause it is. The effects are then deduced from the cause. In the latter case, one begins from the effects. One then postulates the existence of certain kinds of causes from noting invariances in observed events.

The new line of enquiry only becomes possible when the presumedly metaphysical bond between reason and the world is broken. A central contributing factor to the break was the radical creationism of religion. As long as the existence of realised beings was considered to be a metaphysical actualisation of rationally apprehensible essences, then no reference to a separate act of creation or a creative will was necessary in accounting for the existence of beings. Because the intellect grasped the actualised essences directly by virtue of the metaphysical identity of intelligibility and reality, the ultimate explanation of all being was found in the identity of reason and the real. With the breakdown of this assumption of identity, beings began to be looked upon as generated – as becoming – or as being presently instantiated. It was then natural to view their bond with their metaphysical foundation as being established in a separate act of creation. The world became created, chosen, which is to say non-necessary. Correspondingly, the observer's attention shifted from ascertaining *what has happened* to observing *what happens.* This creates an obvious demand for assessing the observer's viewpoint, since from now on it may be thought that the observer's own co-ordinates also come to determine the object of perception.

The view put forward by Scotus and Ockham (discernible already in Augustine) was roughly as follows. Created beings (and the created world in its entirety) because of their contingent nature do not fall under the pure necessity of reason. There is no metaphysical bridge, no participation between reason and the world such as would necessitate the disclosure of reality purely by the intellect. In late medieval nominalism, the world thus becomes for the first time truly foreign and unknown to man: something which can only be made familiar by the aid of systematical observation. At the same time, all the great ideas of reason such as God acquire a new status: even though God is the Creator, he is conceptually independent from creatures. Thus, it is logically possible to question the relation between God and nature and to re-interpret it. The same possibility of doubt applies to the knowability of created beings. All our conceptions of created beings are merely approximations, arrived at within the limits of human epistemic capacity. Reality may be approached without limit, but 'in itself' it cannot in the final analysis be reached. Understanding reality 'in itself' by participation, in the absence of the observer's own perspective, is transformed into the irreducible presence of contextuality. Reality is always *interpreted,* never directly *participated.* Interest then becomes directed at the conditions determining interpretation. Are these invariable in any way?

The centuries-old covenant between metaphysics (ontology) and epistemology was irrevocably placed in a new light at the dawn of modernity. The *transcendentality* of our grasp of reality (its reliance on the tools of thought) becomes one of the conceptual presuppositions of modern man. Obviously, Kant played an important role in the breakthrough to this kind of thinking. He does not, however, have the last word on the subject, because the following problem then presents itself: in what connection and in what way can the idea of transcendentality be made more precise? To what extent can, e.g., the conditions of thought (those conditioning interpretation and understanding) be made clear? No conclusive answer has been readily forthcoming. Consequently, speculation concerning transcendentality has after Kant been a sort of 'free-for-all', and its metaphysical significance could not in the beginning be unambiguously settled. In post-Kantian theology transcendental conceptions of nature led to an idealistic interpretation of the existence of nature *qua* nature. This in turn resulted in a break with the scientific investigation of nature. Partly because of this it has been common for adversaries of transcendentalism to regard many typically transcendental questions as themselves transcendent, and thus to push them into the realm of either speculative or evaluative enquiry. The former, as unresolveable, are easily excluded from the domain of scientific and theoretical reasoning, while the latter have been regarded as belonging to the realm of practical reason.

In light of the aforementioned developments, the relation between scientific and theological modes of understanding has become the subject of wide attention. Because the notion of nature as created has been considered a paradigmatic example of a scientifically unresolveable question (*viz.* Stanley Jaki's paper in this volume), the concept of creation has often been regarded as being at best only an ideological notion inherent in our culture. It has not been considered a viable threat to the autonomy of either theoretical or practical reason. These straitjackets have, however, lately begun to come apart at the seams. This is true also concerning the notion of createdness. This is due not only to views emphasizing the many-facetedness and game-like nature of linguistic signification, but also, and most importantly, to certain systematic results in philosophical logic. These results are fundamentally connected with the notion of transcendentality, although in a much deeper sense than what we find in Kant. Because nature evidently cannot ever be grasped 'purely in itself', ie., as completely rarefied from the transcendental features of the language used in discussing it, it is necessary to acknowledge the transcendental nature of many theological and cosmological modes of speech. Createdness in this case does not have to be treated as an ideology: it is, rather, a transcendental condition of a certain way of using language.

2. Nature in the Grand Project of Western Theology

Although the evolution of theology is in many ways intertwined with philosophy, it has also had its own basic program. From Augustine onwards, the focus of theology has been thought to be 'God and the soul and nothing else'. The theology of Luther, for example, accords with this Augustinian outline. In Luther's thought, the question of nature is almost completely subordinated to the question of the place of man and his God-given destiny. In other words, nature in this grand Augustinian theological project is neither autonomous nor does it have intrinsic value, even if already the Genesis account depicts the relation of God and nature as independent of man ('God saw all He had made, and indeed it was very good' (Gen. 1:31), even before man was made).

This scriptural passage alone would have warranted a theological approach to nature which was separate from man, yet value-based. In that case, God would not necessarily have been available to sanction the pursuit of acquiring control over nature or of neutrally (scientifically and objectively) studying it. In actual fact, however, precisely these attitudes have throughout history been defended on theological grounds. For example, modern experimental science has from its inception been theologically motivated. Reading the 'book of nature' (scientific practise) so as to benefit mankind has been viewed as a duty imposed by the other great book, ie. the Bible. Thus nature in theology has for the most part been interpreted as a stage for man's activity, however responsible, and for his cognitive pursuits. Rarely has nature been considered an end to itself.

In consonance with the attitude outlined above, the problem concerning the relation of God and nature in the late 17th century program of natural theology arises as a typically cognitive problem. This program in turn could not have been launched before the natural sciences gained their independence. The investigation of nature had recently acquired a status that was independent and conceptually distinct from theology. Central to this development were the figures of Newton and Descartes. Newton demonstrated that it is possible for us to theoretically investigate nature as a collection of natural laws, that is, as a collection of mathematically formulated functional dependencies (invariances). No reference to teleological or theological factors is needed. By this time, Descartes had already lent the same view philosophical credibility. An essential element in Descartes' thought was the idea that nature is at heart mathematical: mathematical invariances are even ontologically prior to the universe itself (even if they should be regarded as God-dependent). When the world could be explained by means of invariances heedless of God, a counterbalancing need arose to show that reference to God is nevertheless contained in the invariances. The Enlightenment program of natural theology is as an attempt to address precisely this new problem: With regard to the mathematical laws of nature, are there any features in nature such as would enable us to say something about its ultimate grounds? From here, the notion that both the regularity of nature and some of its *special* features refer to God was but a short step away. The harmony or beauty of creation could be offered up as such a special feature: 'nature speaks of its Creator'.

3. *Createdness in Contemporary Cosmology*

Theological discussions of nature have often been associated with philosophical cosmology. Cosmological explanations have a long history. From the point of view of *temporality* two models of explanation can be said to address the fundamental options: either (1) the world is eternal with regard to both past and future, or (2) the world has a temporal beginning. As regards the world's ultimate *grounds,* the options are either that (3) the world is more primary than the invariances (laws) it exhibits, or that (4) the invariances are more primary than the world. These basic models are found in different variations and combinations in the history of Western thought. In recent years, alternative (2) has formed the primary basis for theoretical studies, as the most popular cosmological theory has been the so-called Big Bang theory. (One could remark that a similar theory is found already in Grosseteste in the 13th century.) Because this theory contains the conceptually difficult notions of a 'singularity' preceding the Big Bang and of a subsequent 'inflationary period', the old debate about whether a temporal beginning should be interpreted as (2.1.) generation or as (2.2.) creation has acquired a renewed sense of urgency. This debate in turn is intertwined in interesting ways with the question about what the human intellect contributes to models (3) and (4).[1]

It turns out that the basic difficulty of recent cosmology lies in finding a theoretical explanation for the determination of the cosmic invariances. Because the conditions obtaining in a singularity or during the inflationary period cannot be directly observed, there have been attempts to extract information pertaining to them from the later development of the universe. This attempt can be characterized as an attempt to solve the so-called *Selector* problematic: what, in the final analysis, determines the world's coming to be in the way that it does and being

---

[1] J.D. Barrow, *Theories of Everything* (Oxford 1991), 28–29.

the way that it is? It seems necessary to make some preliminary assumptions: either the initial conditions determine the state of the universe or not; either the determining conditions are themselves internally dependent on the universe or not; etc. It is easy to see how notions concerning initial conditions would seem to escape precise empirical definition; the beginning and grounds for the universe are inescapably transcendental. I shall look at these questions in more detail later on.

The contemporary debate has produced a cosmological variant of the attempt first voiced in natural theology to solve the problem of createdness. It is a kind of an inverse echo of the way natural theology formulated the problem earlier. Curiously, the relation of this variation to ancient conceptions is then in turn non-inverse, that is, direct. As I mentioned earlier, contemporary studies have attempted to find a special feature in the universe that would make understandable both the existence of the universe and its nature. This is analoguous to the way in which reason guaranteed the intelligibility of nature in premodern thought. Nowadays the question goes as follows: Can cosmological and physical theory attain the level of becoming an ultimate explanation for the universe? Can the universe be explained solely by reference to intracosmic factors? Those who have answered in the affirmative have been charged with attempting the trick of the good Baron von Münchhausen: of claiming to having pulled themselves up by their own boot straps. The problem is this.

When fundamental physics and cosmology aim at a so-called unified theory to completely explain the universe, such attempts arouse a basic philosophical problem concerning the apparent identity of the explanation and that which is to be explained. Either the unified theory exhaustively explains everything that happens in the universe, in which case nothing explains the theory itself. Or it does not, in which case the universe appears to be inexhaustible with regard to explanation. Many philosophers have ended up thinking that an explanation of the universe as a whole can only be given at a level 'higher' than that whole. Because this 'higher' level nevertheless cannot consist solely of abstract or formal things such as logic but must have some semantic content, a basic question of Kantian metaphysics of knowledge again rises as a central topic for discussion. Is nature as a whole a genuine object, to be explained in the same way as objects belonging to nature are explained? Or should nature as a whole at the same time be taken also as somehow being constituted by human understanding itself? It has become ever clearer that the issue is essentially connected with our view of mathematics as an instrument of our intellect. Is the mathematical conception of nature in the final analysis some kind of (synthetic and a priori) participation, after all, or is it an interpretation (synthetic and a priori) of materials given in perception? It is perhaps appropriate to make note in passing of the Kantian background to the problem and its later development.

4. The Problem Regarding the World as a Whole

According to Kant, nature, or the world, is an object of theoretical reason: furthermore reason itself is a part of nature. The limits of the intelligible world are determined by what we can have representations (*Vorstellungen*) or intuitions (*Anschauungen*) of. To the question of whether there is anything on the other side of the limit Kant answers equivocally: yes and no.[2] Kant speaks of 'things in themselves' (*Dinge an sich*): but on the one hand he gives the impression that these 'things' are no separate type of otherworldly and unknowable entities, while on the other hand he gives the impression that this is precisely what they are.

The ambivalence has quite drastic consequences. If there is nothing which is cognitively inaccessible, then 'things in themselves' are merely limit values to our ordinary experiential knowledge, things as they are when they are not objects of our knowledge. In this case there is nothing in existence from which we cannot in principle acquire some experiential information. The fundamental limits to knowledge are at the same time the limits of reality (the world), and beyond these limits there is nothing. The human apparatus governing representations and intuitions is applicable to everything that exists, also to God (if He exists). If on the other hand 'things in themselves' are truly transcendent and not only transcendental, then it is possible that in the noumenal world – corresponding to transcendence – there exist beings which we cannot even in principle access cognitively. In fact, it is possible that all beings are not only transcendental, but also transcendent.

Kant's metaphysics of knowledge allows for the fact that the mind attempts judgements transcending the cognitive domain. Representative of such judgements are certain cosmological and God-related statements. These claims are not cognitive, but regulative. Regulative claims do not communicate knowledge, but emotions. Kant's conception both resembles Christian Neoplatonism (e.g., Pseudo-Dionysius) and differs from it. The resemblance is apparent in that regulative statements concerning God are feelings (*Gefühle*) of some sort. These emotions do not qualify as experiential evidence either in Kant or in Pseudo-Dionysius. They have one epistemological role to play: in taking as their aim the unknowable they disclose the metaphysical limits of knowledge. The difference between the two views, meanwhile, seems to lie with notions concerning the semantics of religious language. According to Pseudo-Dionysius, the signification of a religious utterance always contains as its negation an overcoming of the literal meaning. According to Dionysius we simply 'feel' that none of our expressions suffice to describe the divine reality which is the object. The only thing left is our basic metaphor for divinity: 'darkness'.

---

2   See J. Hintikka, 'Transsendentaalitiedon paradoksi,' *Ajatus* 40 (1983), 20–48.

Kant does not put his trust in metaphors. For example, for him the antinomies are genuine incongruities stemming from a violation of the limits of knowability. They are no mere semantical conundrums which could be explained by pointing to the metaphorical nature of our utterances. Religious expressions are according to Kant mystical in a way that violates the categories of experience. Are all religious expressions such? Again we face Kant's ambivalence. Religious experiences can be interpreted either as transcendental or as transcendent. The latter must simply be rejected as cognitively misleading emotions and false mysticism. The former can be accepted as mystical 'feelings'. These do not denote an actual experience of the world but rather a kind of *metaexperience* – an experience of the limit of the world.

Mystical 'feelings' about nature as a whole must thus necessarily be interpreted as transcendental in so far as they are attempts at an explanation. They are, so to say, limits to explanation which cannot at the same time be counted among the things explained. Although Kant generally seems to deny religious expressions the status of 'metaexperiences', that interpretation becomes possible in his later critical philosophy (surprisingly enough). Pulling yourself up by the boot straps then seems to succeed only if the same straps, ie., the mystical feelings of religion, are firmly tied to a tree-branch growing on a 'higher level', the boundary framing the transcendental conditions of knowledge.

It is thus possible to interpret religious expressions and religious feelings in a way that transforms the mysticism manifest in them into the regulative use of reason: and this regulative function is part of our nature as rational creatures. Two special feelings express mysticism transformed into regulativity: our consciousness of the moral law, and our consciousness of the universe as a whole. It is precisely these two feelings that according to Kant determine our attitude to nature as created; as attitudes they orientate our practical life. Whether these feelings are for Kant in fact at all particularly religious is to an extent left unresolved. In any case, they determine the limits to the mathematical understanding of nature. Schleiermacher, in contradistinction, wanted to make precisely the feeling of the world as a whole into the essence of religion.

5. *The Mystical Nature of the Universe*

Schleiermacher, too, took as his starting-point the idea – ancient in itself – of createdness being the ultimate dependence of being on its grounds. But he believed that the dependence could be experienced as a *feeling* of dependence. The metaphysical content of createdness thus corresponds both to a feeling (*Gefühl*) and to an intuition (*Anschauung*) on the observer's part. The essence of religion is therefore a special mystical experience of reality as a whole. Contrary to Kant,

Schleiermacher thought that the mystical experience is a genuine experience, because its basis is in intuition. Despite this, the object of experience cannot be described, because it is not exhausted in anything observed. This experience manifests itself in various ways (e.g., in different religions) although it is at heart one and the same. Schleiermacher's theory is unclear on this point, but he is still the father of the modern concept of mysticism. Mystical experience is, according to him, man's genuine experience of the 'ineffable' (*ineffabilis*). This kind of experience, understood as the basis for religion, forms a category separate from nature in man's consciousness. For this reason, religious experience and mundane experience are oriented in entirely different directions and do not compete with each other. The idea of the incommensurability of different types of experience has been widely influential in the subsequent philosophy of religion (consider Nygren).[3]

It is perhaps of some interest that the views of Kant and Schleiermacher in one regard are not so distant from the standard views of the scholastics. The relationship between God and nature, the relationship between the *causa prima* (God) and the secondary causes (nature), is viewed in both as involving the limits of knowledge. It is mystical in that sense. What is new in Kant is that createdness can no longer belong to the issues investigated in natural science. This change is due to a change in how the structure of the cognitive faculty itself is viewed. Createdness involves not theoretical, but practical reason.

Whether the notion of mystical 'experience' has any epistemological significance (e.g. regarding the existence of God) is still a question on which completely contrary views are held in contemporary philosophy of religion. It will have to go undiscussed in the present context, except to briefly note one crucial point. If mystical experience concerns the 'ineffable' then the concept of experience itself is apparently rendered problematic. Experience is always intentionally directed at *something*. If we cannot say anything about what it is directed at, then this is equivalent to someone saying 'I have an experience now' without reference to what the experience concerns. Perhaps this is what has led many people to claim with Wittgenstein that mystical experience only points to the fact that we *are* dependent and not so much to the thing *on which* we are dependent. This is to say that mystical experience concerns the existence of a limit bordering nature and a possible transcendence (God), not knowledge of what lies on the other side. Consciousness of a limit creates a feeling that suggests that the solution to the puzzle concerning nature is 'outside the world', as Wittgenstein said. We encounter the very same puzzle when we approach the understanding of nature from the side of mathematics.

---

3   F. Schleiermacher, *Über die Religion: Reden an die Gebildeten unter ihren Verächtern* (Hamburg 1961).

6. Theories of Everything

The Finnish professor of physics Keijo Kajantie has recently described his views on the possibility of a unified cosmological theory. Kajantie comments on J. Horgan's remark about an end to great visions in physics as follows:

> In light of what has been said these kinds of opinions are astonishing. What could be a greater vision than the expectation of understanding the universe in its entirety, from end to end, from the very first moment to the present and future, without in any way sacrificing the basic principles of empirical natural science?[4]

Kajantie's view appears astonishing philosophically, even if we should not welcome an end to cosmological visions as does Horgan. What is astonishing is not so much Kajantie's brash confidence in the possibilities of empirical science making reality intelligible (this is in itself defensible). Rather, it is the fact that his claims betray a nearly total ignorance of the *conceptual* problems involved in the notion of a unified cosmological theory. In what follows I will consider a few of these problems. The largest involves the question about the nature of mathematically 'understanding' the mystery of the universe.

The earlier references to Kant should have sufficed to show that a cosmological unified theory is unavoidably of another sort from other scientific theories. It is, as the saying goes, a 'Theory of Everything'. For this reason it raises questions which are of a metaphysical rather than of a scientific nature. One of these arises right at the outset: what is the ontology of a unified theory? Does it allow for the existence of, e.g., mental entities, and are these supposed to be reducible to so-called 'material' entities? The ontological issue does not, of course, have to be resolved at the outset in constructing a unified cosmological theory, because the resolution may be the conclusion rather than the starting point. But this fact does not negate the existence of the problem: what we assert as existing is always a *theoretical* choice, whether it is made at the outset or at the conclusion of the theory's construction.

This remark is relatively trivial. It can, after all, be said that it is also a theoretical (philosophical) choice to say that existence cannot be exhaustively explained by means of empirical causal chains. But there are much more precise problems which are offshoots of the aforementioned remark. Some of these have to deal with the theoretical constructs of physics and cosmology themselves, while some are of a metaphysical or epistemological nature. Among the former are questions of, e.g., natural laws, initial conditions, the nature of basic forces of nature and

---

4   K. Kajantie, 'Kosmologia B: Uusi täsmätiede,' *Tieteessä tapahtuu* (3)1998, 27–30, 30.

fundamental particles, cosmic invariances, and the symmetry of nature. The latter involve questions regarding the general structure of the universe, the determination of the initial conditions, and the use of cognitive categories.[5]

Talk of the whole of nature or 'the universe' requires equal consideration of three aspects. First, we must be conscious of the conditions of observation itself; second, we must investigate the structure of the universe; and third, we must construct a theory of the contents of the universe, ie., the properties of matter. There is partial overlap, but the question concerning the role of mathematics is central to all three aspects. The first aspect involves philosophical questions concerning the mathematical intelligibility of reality (e.g., the ontology, perfection, and adequacy of mathematics); the second aspect is essentially the point of view of the general theory of relativity; and the third, that of quantum mechanics. The role played by mathematics becomes apparent when we note how our unified theory ought to be expressible in a single formula which would describe the invariance most fundamental to the whole universe – the so-called symmetry relation, along with its conditions, ascribable to the universe as a whole. I will not in the present context discuss the kinds of asymmetries found by physicists and cosmologists and the mathematical difficulties they pose (e.g. chaos and nonlinearity). It is more important to acknowledge the problem we are left with even when we do perhaps gain a unified theory of some sort. First, however, the question about mathematics is worth placing in its historical setting.

Ever since Plato's day, mathematics has been the object of mystification. It has for instance been supposed that mathematical sentences are truths describing a special mathematical reality, and that this reality is at base more fundamental than physical reality or that physical reality is actually one with that mathematical reality. The two are tied together by binding the mathematical and empirical (semantical) content of certain concepts. In contemporary discussion, concepts susceptible to such an intertwinement have been, e.g., 'wave function', 'space-time', 'initial conditions', and 'natural laws'. The semantical operation gains intuitive support from the fact that mathematical descriptions of beings and their behaviour are believed to be *inherent* properties of beings. From this it follows that a certain mathematical sentence is thought to describe reality better or more genuinely than, for instance, a linguistic description does (in these contexts, natural language is often regarded as metaphorical and mathematical language as literal). The shift to intelligibility as participation lies close at hand here. This way of thinking resembles the notion common before the birth of modern science that reality and logic are ontologically one. This fallacy was overcome in the late Middle Ages, but it led to another: the identification of mathematical reality with the physical. The development of the problem can be followed together with the

---

5   Barrow, *Theories of Everything*, 4.

evolution of the general theory of relativity and the subsequent dream of a unified theory.

Kant inherited from Leibniz the problem which is known as the substantivality problem in contemporary discussion concerning the philosophical interpretation of the general theory of relativity.[6] The question is whether the space-time continuum can be regarded as an existent substance with certain properties irreducible to its material properties, and whether the continuum can exist even if there were no material world. Newton had answered this question in the affirmative, since for him time and space were absolutes. Leibniz answered it in the negative, because his metaphysics required that substances consist of relations. Kant's answer acknowledges transcendentality; space-time cannot be studied as a separate whole in a way reminiscent of beings but only as an a priori type of category for physical ('mathematical') reality. When Einstein chose to rely on Kant, the price turned out to be high (as will be seen shortly).

Several kinds of metaphysical preconceptions may accompany a substantivalist outlook. One central supposition is brought to light by asking the following question. If we suppose that there is a unified theory that makes the universe completely intelligible and truly 'saves the phenomena', how strong mathematical realism or Platonism does its mathematics require? It is worth working out how the Platonic version of substantivalism offers an example of conceiving of space-time as an inherently mathematical substance.

7. Infinity and the Universe

Let us consider some isolated examples of mathematical concepts that have been regarded as especially fruitful in physics and cosmology: for instance concepts of spatial geometry such as 'Hilbert space', 'group', or 'the manifold'. The utility of these concepts is based on the fact that by aid of the mathematical tools affiliated with them certain new concepts can be defined, for instance the concept 'string', which in turn is useful in the construction of physical and cosmological theory. 'String' is a useful concept, because it can be ontologically identified as a basic element of physical reality without creating such perceived drawbacks or problems as follow, for instance, from the acceptance of a 'point' as a basic element. Where does Platonic mysticism sneak in? This happens, e.g., when one attempts to take a position in regard to the alternative ways of understanding the interrelations between mathematical and physical reality.

---

6 Cf., e.g., C. Hoefer, 'The Metaphysics of Space-Time Substantivalism,' *The Journal of Philosophy* 93 (1996), 5–27; and cp. with Barrow's models (3) and (4).

From the point of view of a unified cosmological theory it appears to be significant whether the universe is by nature discrete or continuous. Both alternatives have mathematical counterparts: the mathematics of a discrete universe is based on the standard axioms of set theory, whereas a continuous reality appears to require that the so-called continuum hypothesis be accepted – a hypothesis which has in turn been proven independent of other axioms of set theory. On the mathematical level the question then is this: Is there a special kind of infinity between the infinity of the natural numbers (Aleph zero) and the infinity of real numbers (Aleph one)? To the Platonist the situation is intolerable, because he is asking of mathematics an answer to the question about the nature of reality; and the axioms of mathematics cannot be arbitrary, since they form the basis of reality. What is at stake here for the Platonist is the ultimate nature of the universe. If the continuum hypothesis is correct, then reality is inherently continuous, if not, then the basic elements of reality are point-like atoms, logically independent of each other. The question clearly troubled Einstein.

One need not of course accept this way of formulating the question, although it is a philosophical problem. One can deny this, because it is always possible that a mathematical model is again found which can be interpreted as describing actual reality. It is therefore inherently difficult to clearly distinguish between a reference to actual reality and a mathematical construction. Still, the question of whether reality is continuous or discrete is typically transcendental, contrary to what the Platonist believes. Both physical alternatives have their own, logically possible mathematical alternatives: and there is no possibility of choosing between them on the basis of any feature in the actual world. The same problem was posed in post-Kantian spatial geometry with regard to the parallel axiom of Euclidean geometry. Many physicists were willing to swear by the fact that reality is Euclidean until the usefulness of non-Euclidean geometry became evident (with the advent of the theory of relativity).

To sum up: What does the inclusion of the continuum hypothesis in mathematics have to do with real facts about nature? According to the Platonist, everything. Many Platonist representatives of theoretical physics hope to squeeze real information concerning the universe from basic concepts of mathematics, as did those among their predecessors who relied on Euclid. For them, the reality of nature contains inherently mathematical properties. Einstein was in fact drawn to the same view.

This leads us to examine more closely Einstein's dream of a unified theory and, more especially, the metaphysical presuppositions and interpretations it involves. How strong was the Platonism of Einstein's idealised mathematics? Or should we rather regard him as a Kantian transcendentalist?

8. The Theory of Relativity and Mathematics

After he had abandoned the Machian project of eliminating time and space Einstein in the 1920's held that the general theory of relativity lends even more physical weight to time and space than does Newton's theory. Time and space not only determine the behaviour of material bodies but are themselves conditioned by those same bodies. An essential element of Einstein's vision was the view that the mathematical functions which describe the gravitational field so closely govern physical reality (Minkowski's space-time) that they at the same time describe those mathematical instruments (the topological and metrical structural features of the so-called manifold) by whose means the qualitative aspects of physical reality can be described. This is what Einstein emphasizes when he later says:

> If we imagine the gravitational field, ie. the functions g_{ik} to be erased, then not only will no [Minkowsky-type] space be left, but absolutely *nothing*, not even any kind of 'topological space'.[7]

Mathematics thus metaphysically circumscribes reality. The idea has far-reaching consequences. Many contemporary physicists have consequently tried to interpret physical space-time as a so-called 'metric field', by whose aid space-time would be made both real and self-explanatory. The idea is that the model of general relativity describing a certain physically possible world is a three-membered set 'M, g, T'. 'M' is here a four-dimensional, continually self-differentiating and topologically defined collection of points; 'g' is the so-called metric tensor defining the metric and geometric properties of points; whereas 'T' represents the matter and energy of space-time. However useful this kind of model may be in physics and cosmology, philosophically it is obscure. It is something like a substantivalism bought with the coin of Platonist mathematics.

The fundamental philosophical problem pertaining to the mathematics of such a model appears to be the following. The universe is taken as a continuum whose elements are mathematically mutually dependent. But then it is (falsely) assumed that the elements are also logically co-dependent. The problem lies in moving from the mathematical ties between the elements to their logical counterparts. This movement disregards the fact that when what is at issue is a contingent universe, then there may exist mathematical ties between units that are nonetheless logically independent of each other. The problem can be clarified by asking: What additional supposition or premise would justify viewing the elements of the unified mathematical theory as logically interdependent rather than independ-

---

7   A. Einstein, *Relativity: The Special and General Theory* (New York 1961), 155.

ent? Einstein's answer is made very clear in his Leiden lectures from the 1920's. Einstein remarks that

> I cannot imagine a unified and reasonable theory which explicitly contains a number which *the whim of the Creator might just as well have chosen differently,* whereby a qualitatively different lawfulness of the world would have resulted... A theory which in its fundamental equations explicitly contains a constant [of Nature] would have to be somehow constructed from bits and pieces which are logically independent of each other; but I am confident that this world is not such that so ugly a construction is needed for its theoretical comprehension.[8]

What Einstein says is that if a unified theory should be found, then the cosmological constant it contains cannot be arbitrary. It cannot bind together logically independent parts in a random manner. This premise adds to the mathematical tools required in a unified theory, to its 'metric field', a strong Platonic additional premise.

Einstein later reveals this train of thought when he refers to a non-arbitrary, logical double bind between the elements of the unified theory, one which is grasped in 'pure thought': 'In a certain sense, therefore, I hold it true that pure thought can grasp reality, as the ancients dreamed.'[9] This means that contingent facts are mathematically bound together in such a way that it seems impossible to regard the resulting unified theory as random. The shift to logical necessity is accomplished by identifying mathematical elegance with the absolute explanatory force of a theory, that is, its ability to make reality exhaustively intelligible to reason. This quite clearly brings out the Platonic notion often found in conjunction with the dream of a unified theory: a theory is self-explanatory only if it has a necessary nature of a kind recognised by 'pure reason'. The fundamental constant of nature, the so-called cosmological constant, is then at once both a mathematical theorem and an axiomatic metaphysical necessity. Other constants derived from this axiom are theorems which fulfill the symmetry requirement of the universe. In this way they form an explanation for the existence of the entire universe. No outside explanation is needed. This notion was inherent already in the ancient idea of the *logos*. For this reason, a mathematical 'Theory of Everything' can be thought to be both all-explaining by nature and self-explanatory as a product of pure reason.

This was Einstein's dream. It has been looked upon in various ways. Inspired by Kant, Einstein himself dreamed of the kind of axiomatization and deducibility

---

8   Barrow, *Theories of Everything,* 89 (italics added).
9   Barrow, *Theories of Everything,* 180.

outlined above. This path, however, leads to the hidden minefields of idealism in Kant's philosophy. Planck, by contrast favoured an inductive approach in which extramental reality and the categories of reason are kept apart. But in that case the whole dream of a unified theory is relinquished. There is a third way: according to the transcendental semanticist, reality and the theoretical tools required for its grasping cannot (and need not) be kept apart, but those tools are refinable and redefinable without limit.

9. *Unified Theory and Mathematical Illusion*

Now we can perhaps understand what the real philosophical alternatives are as regards the interrelation of mathematics and the pursuit of a unified theory. We can hold two different views. Either the universe is inherently mathematical, in which case the fundamental formula which is the goal of a unified theory discloses the ultimate metaphysical nature of reality; or the fundamental formula is a mathematical formulation of the most general features of physical reality given on a *metalevel,* in which case no metaphysical or *a priori* commitments regarding the mathematical intelligibility of reality are made. The latter approach is that of the transcendentalist.

Those cosmologists pursuing Einstein's dream presuppose a metaphysical leap: the ultimate formula discloses the ultimate explanation of reality only on the supposition that the fundamental formula is a theorem of pure mathematics, that is, if reality is inherently and ontologically mathematical. This leap can be viewed as exemplifying the fact that reference to mathematics as the grounds for the world being as it is in any metaphysical sense is a sign of conceptual confusion. There are good reasons to claim instead, as did John Duns Scotus, that an inherent feature of the world is that it could have been other than it is. Furthermore, should it have been other than it is, it could still well have been mathematically approachable. Mathematics in itself does not dictate any cosmological necessity.

What is essential is that when a physicist or a cosmologist wishes to apply mathematical tools to some concrete physical object, the object studied may itself display features which can enrich the mathematical 'toolbox'. Just because of this fact it is often difficult to say when a model abstracted from reality enriches mathematical thought and when mathematical thought is applied to create a model of reality.[10] This reciprocal influence is just what creates the illusion of identity between mathematics and physical reality. In truth what is disclosed here is not ontological identity but the efficacy of mathematics as an instrument of description and explanation. (For example, the use of set theory, left ontologically unin-

---

10 Cf. with Barrow, *Theories of Everything,* 182–99.

terpreted, in the investigation of the symmetrical properties of the universe; or the use of interpreted axiomatic, say, Euclidean, geometry in physics.) Therefore, it is not true that mathematics (e.g. set theory) necessarily speaks of itself when it is used to describe physical reality. For example, the nonlinearity of a chaotic reality does not constitute a counterexample to the use of mathematics in physics. Instead it challenges us to develop a new mathematics. The fact that mathematics in its entirety cannot be constructed nominalistically has nothing to do with the mathematical Platonism encountered in cosmology.

10. *Modal Theory and 'Selecting' a World*

Already at the beginning of this paper I referred to the so-called Selector problematic: out of all possible initial conditions, why did certain conditions obtain during the initial inflationary stages of the universe? There are innumerable logically and metaphysically possible universes, and the 'zero' possibility (that of there being no universe at all) is also a logical possibility. Why did precisely a certain universe (or certain universes, if there are many) become realised out of all possible universes? Or should we assume that all possible universes have been realised?[11] In the context of a unified theory the problem takes the form: why is precisely a certain formula realised out of all logically possible ones?

There have been attempts to prove that the constants of nature obtaining in the chaotic early stages of inflation were the only real alternatives in light of the universe's later development.[12] The Selector argument follows the logical form: if p, then not-possible that not-p. 'Not possible' is qualified in Platonist fashion as concerning understanding, reasonability, and ultimate reality. The argument is of course erroneous even as qualified, because it only says that when the fundamental formula of the unified theory has been arrived at, it is understood how the world cannot be other than it is.

In his *Theories of Everything* John Barrow presents an argument which reveals in an interesting fashion the concealed premise concerning the unity of the intellect and the material world. As Barrow formulates it, the question goes as follows: Why were the initial conditions such that they allowed for the development of our current mathematical inquiries? Barrow's tentative answer is that conscious life

---

11 Cf., S. Knuuttila, 'Plenitude, Reason and Value: Old and New in the Metaphysics of Nature,' in: C. Bengt-Pedersen and N. Thomassen (eds.), *Nature and Lifeworld: Theoretical and Practical Metaphysics* (Odense 1998), 139–51. For an assessment of the many-universes theory see J. Leslie, *Universes* (London 1989).
12 It may be noted that the 'new exact science' described by Kajantie has the same goal. It pins its hopes on new possibilities of measuring anomalies in the background radiation of the universe.

and its attendant mathematical faculties are a special feature of the universe, and that the actualisation of this feature actually interconnects with the way the initial conditions were determined. The initial conditions had to be specific enough to allow as the end product of several billions of years of evolution the mathematical description and explanation of the universe. The initial conditions were so to say preset to enable the universe's algorithmic compressibility. By 'algorithmic compressibility' Barrow means the ability of the conscious mind to describe reality by means of rules or formulas, the most efficient of these being mathematics. Mathematics, then, is not only an independent and transcendental tool for the cosmological and physical explanation of the universe, but a product of the universe for its understanding of itself. The initial conditions are explained by one of the universe's own products: mathematics explains both its own existence and the nature of the universe as a whole.[13]

One can easily see that the argument is circular. The existence of the universe is explained by referring to a special feature of the universe; at the same time, it is presumed that that feature follows from the universe's existence. In addition, the unity of mind and reality is assumed as a further premise. Barrow does not mean to exclude the possibility that the universe might have been mathematically describable or explicable even if the initial conditions had been different. But he does claim that – the initial conditions being exactly what they are – evolution will guide man *qua* mathematician to recognise the obtaining cosmological constants as being somehow the only ones which the human mind can accept as explanatory. Because all natural constants in a unified theory can be reduced to a certain fundamental constant and because this fundamental constant is both mathematical and, from a rational point of view, the only acceptable one (in this sense it is necessary and 'purposive'), it has to be the Selector of all other constants. In a manner resembling ancient thought, it is thought that the mathematical faculty of the human mind directs us to the source which will disclose why the universe is necessarily what it is.

This way of thinking is, however problematic; seen from a modal theoretical point of view, it may even be said to be completely wrong-headed. The intelligibility of the world does not have to rely on any principles shared by all intellects, other than the ever evolving and expanding possibilities of pure logic and pure mathematics. No necessary and metaphysical tie between the intellect and any *one* possible or actual world is required.[14] As I see it, this approach frees in a correct manner the human mind from the constraints of biologism. Even if we do consider ourselves the products of natural evolution, our understanding does not

---

13 Cf., Barrow, *Theories of Everything*, 165.
14 Cf., J. Hintikka, *The Principles of Mathematics Revisited* (Cambridge 1996), 197–98.

have to be restricted to any one or formulaic mundane or natural mode of thinking in the conventional sense.

Kant may serve as a warning example of the kind of transcendentalism that ties our understanding of nature to a single set of forms (the notoriously conservative Kantian categories). By contrast, a transcendentalist approach, which only assumes that the fundamental tools of reason are tied to what is logically or mathematically possible, might approach nature in a fashion more akin to poetry. Such a philosophy draws from the inexhaustible wellsprings of conceptual creativity. Referring to Kotkavirta's study of Kant and Hegel in this volume, one could say that the transcendentalist would be happy with the idea of progress *ad infinitum,* and that he would find Hegel's craving after the Absolute infinite delusional. Let whoever doubts the fruitfulness of this attitude contemplate the following universal human experience: how often we learn to more deeply appreciate our reality by thinking about how things *could be* rather than by thinking about how things are.

As regards theology and its attitude towards science, the approach outlined above entails a critical attitude in all directions. On the one hand, there is no reason to belittle the unfathomable mystery that being and nature pose. Theology certainly must be critical towards modes of thinking which do away with the mystery or ignore it. On the other hand, there is no reason to doubt that our theoretical knowledge will expand decisively and that it will quite possibly completely alter our world-view. It is doubtful, however, that even future scientific world-views will be directly applicable to grand Theist projects. Constructing a theology on a specific metaphysical idea is always dubious. It is healthier to regard everything transcending nature as cognitive 'darkness', as Christian Neoplatonism does, even if the acceptance of a transcendental point of view allows us to ever more closely straddle the borderline between or our knowledge and that 'darkness'. The utility of this kind of critical and doubtful approach may lie in the way it leaves room for more immediate religious feelings and ethical motives. These may encourage a personal attitude towards nature which stems from a theological standpoint rather than a certain theory of being.[15]

---

15 An earlier version of this paper has seen print in Finnish: see 'Teologia, kosmologia ja luonto,' in: S. Nurmi (ed.), *Hurja luonto Abrahamista Einsteiniin* (Helsinki 1999). The article has been revised for this volume. English translation by Taneli Kukkonen.

Notes on Contributors

Kari Enqvist
Professor in Particle Physics at the University of Helsinki. Cosmology Project until 1999 and Theory Program Leader in the Helsinki Institute of Physics since 1999. Chairman for the Nordita Subatomic Physics Committee, 1999–2001. Ph. D. in Elementary Particle Physics at the University of Helsinki, 1983. Research Fellow at CERN, Switzerland, 1984–1986 and at the Physics Department, University of Wisconsin; Madison, 1986–1987. Assistant Professor in Nordita, Copenhagen, 1990–1994. Many research awards: Valtion tiedonjulkistamispalkinto, 1995; Lauri Jäntti foundation award, 1997; The Magnus Ehrnrooth Foundation Physics Award 1997; Tieto-Finlandia, 1999. Publications: 148 scientific publications. Tyhjästä syntynyt (1994) with J. Maalampi; Näkymätön todellisuus (1996); Olemisen porteilla (1998) (Winner of the Annual Tieto-Finlandia Award).

Jaakko Hintikka
Jaakko Hintikka has been Professor of Philosophy at the University of Boston since 1990. Habilitation, 1956, in Philosophy at the University of Helsinki. Professor of Practical Philosophy in Helsinki 1959–1970, Academy Professor in Finland 1970–1981. Memberships and posts in several international scientific societies. Publications include among others: Investigating Witgenstein (1989); The Logic of Epistemology (1989); On the Methology of Linguistic (1991); The Principles of Mathematics Revisited (1996). Editor in Synthese Library since 1965.

Stanley L. Jaki
Distinguished University Professor in Physics and Astronomy at Seton Hall University, South Orange. He was born in Hungary and studied Philosophy, Mathematics, and Theology. Received the doctorate in Theology in Rome, 1950. Has studied in Berkeley, California an at Princeton. Gifford-Lecturer at the University of Edinburgh in 1974–1975 and 1975–1976. Visiting Professor in many Universities in the USA and Europe. Templeton Prize, 1987. Publications: many books on history and the philosophy of Science, also translated into many languages: The Relevance of Physics; The Origin of Science and the Science of its Origin; Science and Creation; Planets and Planetarians: God and the Cosmologists (1989); Is there any Universe? (1992); Meanings to Message: Treatise on Truth (1998).

Tarja Kallio-Tamminen
Tarja Kallio-Tamminen is working as a researcher in the science-theology project (organizer of the 'Infinity, Causality and Determinism' qolloquium) at the University of Helsinki. She completed a degree in Elementary particle physics at the University of Helsinki and later continued her studies in the Department of Philosophy at the same university. She has lectured and written many articles concerning the foundations and interpretations of quantum mechanics in finnish. Her PhD dissertation 'Observers and Actors, the Role of Human Beings Within the Paradigms of Classical and Quantum Physics' is forthcoming.

I. A. Kieseppä
I. A. Kieseppä received a Doctor's degree (PhD) in Philosophy at the Faculty of Arts of the University of Helsinki in 1995. An improved version of his dissertation was subsequently published in the Synthese Library by Kluwer Academic Publishers. He received the position of Docent of Theoretical Philosophy at the Department of Philosophy of the

University of Helsinki in 1997 and is currently (1998–2001) employed as a Post-doc researcher at the Academy of Finland. His current research interests are mostly focused on the philosophy of statistics, but he has also participated in the discussion concerning the interpretation of quantum mechanics and their significance for the philosophy of mind. His publications in this field include 'Is David Bohm's Notion of Active Information Useful in Cognitive Science?' and 'On the Difference between Quantum and Classical Potentials: A Reply to Hiley and Pylkkänen', both published in: P. Pylkkänen, P. Pylkkö, A. Hautamäki (eds.), Brain, Mind and Physics (1997).

Heikki Kirjavainen
Doctor of Theology, Professor in theological ethics and philosophy of religion at the University of Helsinki since 1993. His professional interests include semantics (especially medieval), theories of religious language, philosophical logic and its applications, analysis of ethical argumentation, applied ethics (especially biomedical ethics) and history of theology from an analytical viewpoint.

Olli Koistinen
Olli Koistinen received his Ph.D. 1991. He is professor of theoretical philosophy at the University of Turku. Koistinen is specialized in the metaphysics of early modern philosophy. He has published papers on Spinoza and Descartes in Ratio, Southern Journal of Philosophy and Acta Philosophica Fennica; a joint paper with Arto Repo on Leibniz theory of modality is forthcoming in Studia Leibnitiana. Together with professor John Biro (University of Florida) Koistinen has been editing an anthology on Spinoza which is forthcoming from Oxford University Press. Koistinen's paper in that anthology deals with Spinoza's concept of causation. Koistinen is also interested in analytic metaphysics and action theory. He has written with Arto Repo a paper on metaphysical vagueness which is forthcoming in an anthology edited by Terence Horgan and Matjaz Potrc and which most plausibly will be published by Oxford U.P. This spring (2000) Koistinen has completed a monograph on the theory of action which should see the light of day later this year.

Jussi Kotkavirta
Born 1954. A lecturer of philosophy at the University of Jyväskylä and a docent of philosophy at the University of Joensuu. Current interest of research: Kant and German Idealism; Moral Realism; Philosophy of Personhood; Philosophy of Technology; Philosophical Aesthetics.

Taneli Kukkonen
M.Ph, M.Th. Works as a junior researcher at the Department of Systematic Theology in the University of Helsinki. He is preparing a Ph.D. dissertation on the relationship between cosmology and modal metaphysics in the Arabic philosophical tradition. The project was undertaken in 1998, when the author worked for prof. Simo Knuuttila as a researcher assistant for the Academy of Finland. Publications set to appear shortly include: 'Possible Worlds in the Tahāfut al-tahāfut. Averroes on the Plenitude and Possibility,' Journal of the History of Philosophy 38/3 (2000); 'Possible Worlds in the Tahāfut al-falāsifa. Al-Ghazālī on Contingency and Creation,' JHP 39/1 (2001); 'Proclus on Plenitude and Time,' Dionysius 19 (2000); 'Infinite Power and Plenitude. Two Traditions on the Necessity of the Eternal,' in: J. Inglis, Medieval Philosophy and the Classical Tradition in Islam, Judaism, and Christianity (2000); 'Missing Links in Lovejoy's Great Chain. Diversity and Order in Ancient and Medieval Thought,' in: M. Oksanen & J. Pietarinen (eds.), Philosophy and Biodiversity (2001); 'Al-Ghazālī's and Averroes on the End of All Things' in: Medieval Philosophy and Theology; 'Plenitude, Possibility, and the Limits of Reason,' Journal of the History of Ideas 62/1 (2001).

Raimo Lehti
Raimo Lehti is an emeritus professor of mathematics at Helsinki University of Technology. He obtained a. Ph.D. in University of Helsinki, and worked there 1955–1964 at the Astronomical Observatory. From 1964 to his retirement in 1989 he worked at the Institute of mathematics of Helsinki University of Technology. His mathematical speciality was multilinear algebra; he has also published articles about relativistic astronomy. He has written some books (in Finnish language) and several articles (mostly in Finnish language) about history and Philosophy of mathematical and physical sciences, especially astronomy.

Eeva Martikainen
Professor of Modern Theology at the University of Helsinki. Director of the Research Project 'Theology and Science' at the Academy of Finland. Member of the Scientific Council of the University of Helsinki and Board Member of the Council of the Scientific Society in Helsinki. Dissertation in Theology (1981) and, after promotion, Junior Research Fellow in the Academy of Finland. Publications: 'From Dissonance to Consonance? Discussion Concerning the Relationship of Science and Theology in the Research Project "Physics, Philosophy and Theology" Initiated by the Vatican,' in: Urho Ketvel et alii (eds.), Vastakohtien todellisuus [The reality of contrasts] (1996), 109–116; Metaphysic: A Shared Challenge for Theology and Philosophy (1997); Religion als Werterlebnis (2002).

Juhani Pietarinen
Professor of Practical Philosophy, University of Turku (Finland) since 1973. Doctor of Political Sciences, University of Helsinki(1972; thesis 'Lawlikeness, Analogy and Inductive Logic'). Research on social justice, bioethics, environmental ethics, and history of philosophy(especially Plato and Spinoza). Publications include monographs Lawlikeness, Analogy, and Inductive Logic (North-Holland, 1972); Ilon filosofia. Spinozan käsitys aktiivisesta ihmisestä [Philosophy of Joy. Spinoza's Conception of Active Mind.](Helsinki University Press, 1993); Platonin harmonisen mielen etiikka [Plato's Conception of Harmonious Mind](Helsinki University Press, 1996, and papers on Plato, Hobbes, Spinoza, principles in bioethics, values of nature, biodiversity, and information society. Member of the Administrative Board of the Philosophical Association of Finland (1978–93); Vice President (1978–81); member of the Administrative Board of the Finnish Association for the Philosophy of Law (1990–96); editor of Ajatus (periodical of the Finnish Philosophical Society) 1984–90; member of the Finnish Academy of Science (1987–); member of the Rome Club of Finland (1989–96); member of the Advisory Committee for Ethical Questions of the National Board of Social Affairs and Health Care (1992–1994); Vice-President of the International Institute of Argumentation (Yerevan) (1991–); honorary member of the Armenian Philosophical Association (1991–); Dean of the Faculty of Social Sciences, University of Turku, 1989–199

Sami Pihlström
Sami Pihlström defended his Ph.D. thesis in the Department of Philosophy, University of Helsinki, in 1996. He is Docent of Theoretical Philosophy at the University of Helsinki, Docent of Philosophy at the University of Turku, and Docent of Philosophy, especially Axiology and Philosophical Anthropology, at the University of Kuopio. He currently works as a Post-Doctoral Research Fellow at the Academy of Finland (1999–2002). His main publications include Structuring the World (Acta Philosophica Fennica, 1996, Ph.D. thesis) and Pragmatism and Philosophical Anthropology (1998) as well as several articles and review papers in both international and Finnish philosophical journals and collections. He received in 1997 the Charles S. Peirce Essay Prize from the Charles S. Peirce Society. Dr. Pihlström is, since 1996, one of the editors of Ajatus, the Yearbook of the Philosophical Society of Finland. He

is also one of the authors of a new Finnish introductory textbook on philosophy, Odysseia: Matka filosofiaan (2000). His current research interests lie in the interconnections between the tradition of pragmatism, Kantian transcendental philosophy, and the philosophical issues of realism and idealism.

Juleon M. Schins
Graduation, 1986, at the University of Amsterdam in Experimental and Theoretical Physics. Dissertation, 1992, prepared at FOM-Institute for Atomic and Molecular Physics (Amsterdam) on Photodissociation and Metal-Induced Dissociation of Hydrogen Molecules. Until 1995 at the Commissariat à l'Energie Atomique, Saclay, and Ecole Nationale Supérieure de Techniques Avancées, Palaiseau (France) on Electronic Free-Free Transitions in Intense Laser Fields Using the Auger Effect in Rare Gas Atoms. From 1995 to 2000: assistant professor at the University of Twente (the Netherlands), on Cytometry (Biophysics). From 2000 at the Swammerdam Institute for Life Sciences (University of Amsterdam) on Coherent Anti-Stokes Raman Microscopy and Third Harmonic Generation (nonlinear optics). In the field of philosophy, Schins published two contributions in Mathematical Undecidability, Quantum Nonlocality, and the Question of the Existence of God (eds. A. Driessen and A. Suarez. Dordrecht: Kluwer Academic Publishers 1997), studying the implications of number theory and quantum mechanics for the existence of an external, man-independent world.

Reijo Työrinoja
ThD, Professor in Systematic Theology, Chairman of the Department of Systematic Theology. Other publications in medieval philosophy and theology: 'Regularity of Will and the Problem of Egoism'. Bazán, C., Andújar, E., Sbrocchi, L. G. (eds.), Moral and Political Philosophies in the Middle Ages II Laboratoire de la pensée ancienne et médiévale I, 2 Legas. (1995), 949–963; 'Aureole's Criticique of Henry of Ghent's Lumen medium', in: Jan Aertsen und Andreas Speer (Hrsg.), Was ist Philosophie im Mittelalter? Veröffentlichungen des Thomas-Instituts der Universität zu Köln. Miscellanea Mediaevalia, Band 26 (1998), 622–628; 'Lumen medium. Henry of Ghent on the Accessibility of Theological Truths', in: G. Holmström-Hintikka (Ed.), Medieval Philosophy in: Modern Times, Synthese Library 288 (2000), 161–182; 'Fides et visio. On Visual Metaphysics of Knowledge and Religious Belief in the Middle Ages,' in: Andreas Speer (Hrsg.) Die Dionysius-Rezeption im Mittelalter. Archiv für mittelalterliche Philosophie und Kultur, Heft 6. (2000), 115–131; 'Faith and the Will to Believe. Thomas Aquinas and Robert Holkot on the Voluntary Nature of Religious Belief' Documenti e studi sulla tradizione filosofica medievale, Vol. XII (2001).

Rainer E. Zimmermann
Rainer E. Zimmermann, professor of philosophy in Munich and Kassel is a life member of Clare Hall, Cambridge. Studied physics and mathematics in Berlin and at Imperial College London, then philosophy, history, and literature in Berlin. Holds a PhD both in mathematics and in philosophy. Habilitation in Kassel on natural philosophy (Schelling). The research work deals with interdisciplinary approaches to the interfaces among the sciences, philosophy, and the arts. Author of about 200 papers on these topics, recently published the books Die Rekonstruktion von Raum, Zeit und Materie (1998), The Klymene Principle (1999), Das Drängen des Blickes (2000).

Index

Aquinas, Thomas 19, 31, 45–50, 60
Arabic philosophy 22–23, 48
Aristotle 52, 59–60, 72, 162, 210, 216, 218, 219
   on causality, see causality
   on first philosophy 8, 61–63, 70
   on hylemorphism, see hylemorphism
astronomy 11, 162, 194, 195
atheism 29
Augustine of Hippo 193–194, 199, 228, 230
autonomy 8, 169
   of human being 8
Averroes 20, 25–32, 35–39, 42–43, 48–50, 167
Avicenna 25–27, 29, 33, 40–42, 207

Bâqillânî, al- 40
Berkeley, George 104
Bible 16, 193, 194, 230
Big Bang cosmology 21, 140, 162, 231
Biel, Gabriel 56, 58
Bohm, David 90, 102
Bohr, Niels 7, 90–95, 97–99, 102–105, 162, 197, 206, 208, 209, 211
Born, Max 90, 211

Calvin, Jean 57–58
cause 111, 114, 117
   Aristotelian 9, 10, 111–113
   four causes (Aristotelian) 9, 25–26, 93
   secondary 10, 35, 46–47
causal asymmetry 114–115
causality 9, 13, 14, 92, 93, 95, 97–98, 103, 139, 142, 166, 215, 216
   Aristotelian Concept of 9, 10, 25–27
   divine 19, 45
   efficient 22, 93
   final 38
   immediate 46
   natural 19, 35
   secondary 32, 54
Chardin, Teilhard de 202
coarse graining 121–124, 126, 135, 136, 139
complementarity 209, 219
   see also Quantum mechanics

consciousness 83, 93, 97, 104, 105, 165, 221, 234, 235
Copernicus, Nicholas 58, 162, 193, 195
cosmology 7–9, 15–17, 120, 147, 157, 227, 231, 232, 239
   Christian theological 193–203
creation 20, 21, 22, 23, 25, 27, 28, 34, 35, 36, 42, 43, 45, 48, 55, 56, 80, 200, 228, 229, 231,
   ex nihilo 43, 56
Darwinism 202
Descartes 11, 12, 13, 49, 57, 63–68, 90, 93, 98–99, 103, 107, 216, 218, 230
determinism 39–43, 69, 90, 91, 114, 136–138, 167, 216, 218
dualism 94
   Cartesian 65, 68, 99
   wave-particle 89, 214–215

Einstein, Albert 7, 91, 137, 146, 147–152, 154–158, 161, 174, 195, 238–242
electrodynamics 147, 148, 150, 162
electromagnetism 134, 144, 174
   electromagnetic fields 133
   electromagnetic interaction 14, 134
emergence 119
   Bunge's definition of 121–122
   weak 121, 122, 135–137
emergentism 120
Enlightenment 231
epistemology 11, 13, 15, 117, 149, 207, 208
the EPR-controversy 91
eternity 23, 24, 138, 188–189, 231
ethics 166–168, 191–192, 221–222
evolution 201, 202, 221, 222, 237–238, 244
explanation 92, 112, 113, 115–118, 119, 120, 125, 228, 231, 232, 234, 241, 242, 244

Fârâbi, Abû Nasr al- 39
Fichte, Johann 84, 87, 165
final theory, see Theory of Everything
freedom of will 10, 12, 37–38, 96, 101, 107–110, 166, 169, 191, 219

Galen 21–22

251

Galilei, Galileo 102, 143, 149, 194
geometry 137, 152, 153, 155, 156, 174, 176, 178, 181, 238, 239, 243
Ghazâli, al- 10, 20, 24–30, 32–38, 41–43, 48–49
the Glashow-Weinberg-Salam model 134–135
God 8, 10, 12, 98, 227, 230, 235
  and causality 19, 31, 33, 45–60
  Creator 8, 10, 193, 194, 198–200, 228, 231
  as final cause 9, 26, 46
  as first mover 26, 30
  the existence of 22
  existence of 66–68, 222
  free will of 6, 47–48, 52
  essence of 20, 39, 47–48, 66, 166–167
  omnipotence of 16, 27, 51–52
gravity 3, 138
  Newton's theory of 131
  Quantum gravity theory 140, 174
  see quantum physics
Gödel's incompleteness theorem 201, 219–220

Hegel, Georg Wilhelm Friedrich 12, 74, 83–87, 165, 245
Heisenberg, Werner 92–93, 127, 141, 162, 210, 216
holistic
  scientific demonstration 6
Holkot, Robert 54–55
Hume, David 33
hylemorphism 207, 216, 218–223
Hölderlin, Friedrich 83, 165

infinity 10, 12, 73–87
interaction (physics) 7, 129, 134, 135, 140, 197, 213–214
  weak 7, 138
  strong 7
  electromagnetic 7
  gravitational 7

Jacobi, Friedrich Heinrich 84–87

Kant, Immanuel 12–13, 63, 68–69, 83, 84, 86, 87, 103, 208, 217–219, 229, 233–236, 238–239, 241–242, 245
  on the antinomes 12, 73, 77–79
  on infinity 72–82
  philosophy of 12–13, 102

on moral law 75
on synthetic a priori knowledge 61, 68, 76
on transcendental philosophy 68, 75, 103–107
knowledge 35, 39, 62–63, 72, 73, 75, 77, 78, 80, 83, 136, 162, 166, 171, 191, 192, 201, 232, 233, 234, 245
  a priori 68
  rational 12
  scientific 185, 197–198, 207, 210, 223

Laplace, Pierre-Simon de 216
laws
  mathematical 11
  of nature 34, 150, 161
Leibniz, Gottfried Wilhelm 68, 170, 201, 223, 238
logic 14, 34, 71, 114, 117–118, 169, 197, 201, 202, 219, 229, 232, 237, 244
Lorentz, Hendrik Antoon 145–147, 149–151, 161
Luther, Martin 56–58

Mach, Ernst 154, 158
materialism 24, 120, 190
mathematics 94, 219–220, 232, 234, 235, 239, 240, 242, 244
Maxwell, James Clerk 162
Maxwell's equations 144–145, 149
  on electromagnetism 144
measurement problem 13, 14, 90–91, 96, 125–130, 131
  see quamtum mechanics
mechanics
  Newtonian 11, 89, 131, 152
  see also quantum mechanics
Melanchton, Philipp 59
metaphysics 7–13, 15, 17, 38, 41, 61, 62, 63, 68, 69, 70, 73, 77, 90, 94, 96, 113, 194, 205, 227, 232, 233, 238
Michelson, Albert Abraham 144
Minkowski Hermann, 151–153, 161, 162

natural
  forces 14
  sciencies 8, 11
  theology 11
naturalism 32, 167–168
nature 11, 12, 19, 27, 33–36, 38, 43, 45, 47, 49, 52, 53, 57–60, 62, 167, 73, 74, 75, 81, 82,

89, 92, 99, 101, 106, 133, 137–138, 150, 154, 166, 167, 190, 207, 227, 228–239, 245
necessity 39, 71, 167
neoplatonism 233
neuroscience 120
Newton, Isaac 7, 102, 143, 149, 158, 162, 175, 230, 240
nominalism 10, 53, 228
normal science, see philosophy of science

observer 147, 158, 207
  in classical physics 90–91, 207, 208
  in quantum mechanics 90–91, 93, 94
occasionalism 31–36, 41, 48–50
Ockham, William 10, 47, 52–56, 227–228
ontology 9, 10, 47, 55, 229
  Aristotelian 9, 70, 72
  Hegelian 83
  in contemporary physics 90, 93, 94, 102, 236, 237, 238
Osiander, Andreas 11

particle physics
  elementary particles 7
Pauli, Wolfgang 93
Penrose, R. 178, 180, 181, 182, 219, 220
percolation theory 184–185
Philo 23, 40
Philoponus, John 27–29
philosophy 12, 16, 19, 24, 84, 120, 132, 169, 185, 200, 202, 205, 211, 217, 230, 238
  Aristotelian 9, 12, 23, 25, 27, 29, 31, 61–62, 70, 93
  contemporary 7
  moral, see ethics
  'first philosophy' 4, 61–63, 70
  of quantum mechanics 103, 105, 206, 216
  of science 7, 90, 102, 105, 161, 197
physics 7, 61, 63, 64, 66–70, 72, 212, 232
  classical 89, 90–95, 97–98, 137
  contemporary 7, 8, 12
  history of 7–8, 89
  modern 106
  Newtonian 89
  quantum, see quantum physics
Planck, Max 91
  Planck mass 140, 141
  Planck time 140
Plato 21–22, 25, 59, 237

positivism 7
postmodernism 7
post-Wittgensteinian philosophy 7
problem of measurement 96, 122, 125–131, 212, 213
Pseudo-Dionysius 233

quantum computation 184–185
quantum field theory 134, 135, 137, 138, 141
  see also quantum physics
quantum mechanics 89, 90, 92, 119, 122, 127, 130, 205, 206, 207, 208, 209, 210, 212, 215, 216, 218
  standard interpretation of 13
  see also quantum physics
quantum physics 13, 139
  Copenhagen interpretation of 13, 90–100, 101–105, 110, 197, 206, 216
  probabilistic nature of 93, 125–126, 128, 139
Quantum Chromo Dynamics (QCD) 135
quantum gravity 165, 174, 179, 187, 188
  loop theory of 174, 176, 177, 180
  superstring theory of 174
  topological aspects of (TQFT) 181, 182, 183
quantum theory 13, 14, 176
Quine, Willard Van Orman 71

realism 8, 9, 105
  Aristotelian 8, 9
reality 13, 89, 94–99, 101, 123, 130, 135–137, 142, 228–229, 237
reductionism 90, 91, 194
Reinhold, Erasmus 84
relativism 7
relativity, Einstein's
  General Theory of 137–142, 155–158, 174
  Special Theory of 137, 138, 147–149, 155–158
  principle of 143, 144, 147, 148, 149, 156–157, 158, 161–163
religion 68, 73, 83, 110, 227, 234, 235

science
  and explanation 92
  modern 7
  natural 92
  practice of 12
Schelling, Friedrich 83, 165

Schleiermacher, Friedrich 234–235
Schrödinger, Erwin
   "Schrödinger's cat" 128, 129, 212–213
   Schrödinger equation 127–129, 208, 210, 213, 215, 218
Scotus, John Duns 52, 227–228, 242
semantics 234
space 12, 14, 133, 137, 144, 147–149, 154
   Hilbert space 177, 179, 182, 238
   Minkowski space 151–153
   space-time 12, 91, 94, 131, 137, 139–142, 151, 156, 169, 170, 175, 176, 179, 180, 187, 237
spin theory
   spin foam 181, 182
   the spin geometry theorem 178
   spin network 177, 178, 179, 180, 181, 182, 183, 184
   see also superstring theory
   see also Theory of Everything;
   see also M-Theory
Spinoza, Baruch de 16, 68, 72, 74, 85–86, 165–172, 174, 187–192
Stoics 24, 166
subject-object structure 91, 103–104
substantial forms 45–47

Sufficient Reason, principle of 38, 167
symmetry 99

T-symmetry 138
T-violation 138
teleology 11, 46
theologians 5, 13
theology 7–9, 11–13, 15–17, 23, 56–58, 62–63, 194, 198, 200–203, 219, 223–224, 229, 230–232, 245
   Ash^{c}arite 26, 36, 38–39
Theory of Everything (TOE) 14, 31, 61, 122, 132, 137, 140, 142, 165, 174, 175, 187, 232, 236–238, 243–244
time 12, 14, 27–30, 37, 50, 133, 137, 149, 188–189

uncertainty principle 127–128, 141
unified theory see Theory of Everything

via moderna 51, 56

wave function 13, 126–128, 205, 206, 208, 209, 210, 211, 212, 224–226, 237
will see freedom of will
Wittgenstein, Ludwig 235

Contributions to Philosophical Theology

Edited by Gijsbert van den Brink, Vincent Brümmer and Marcel Sarot

Vol. 1   Gijsbert van den Brink / Marcel Sarot (eds.): Understanding the Attributes of God. 1999.

Vol. 2   Marcel Sarot / Gijsbert van den Brink (eds.): Identity and Change in the Christian Tradition. 1999.

Vol. 3   Marcel Sarot: Living a Good Life in Spite of Evil. 1999.

Vol. 4   William Hasker / David Basinger / Eef Dekker (eds.): Middle Knowledge. Theory and Applications. 2000.

Vol. 5   Wybren de Jong: Identities of Christian Traditions. An Alternative for Essentialism. 2000.

Vol. 6   Eeva Martikainen (ed.): Infinity, Causality and Determinism. Cosmological Enterprises and their Preconditions. 2002.

Werner Ustorf

Sailing on the Next Tide

Missions, Missiology, and the Third Reich

Frankfurt/M., Berlin, Bern, Bruxelles, New York, Oxford, Wien, 2000. 274 pp.
Studies in the Intercultural History of Christianity. Edited by Richard Friedli,
Jan A. B. Jongeneel, Klaus Koschorke, Theo Sundermeier and Werner Ustorf.
Vol. 125
ISBN 3-631-37060-1 · pb. € 37.80*
US-ISBN 0-8204-4815-X

When German missiologists started to *re-import* their dream of a dominant Christianity to central Europe, there were more similarities between the missionary and the national socialist utopias than the post-war consensus would like to admit. Fascism to many missiologists became the desired breaking point of modernity, a revival of the *Volk's* deep emotions and a breakthrough of the archaic spirituality they had long been waiting for. Upon this tide they wanted to sail and conquer new territories for Christ. This study, therefore, will address the issue of mission and Nazism primarily in the light of the struggle of Christianity for a place or a home within and vis-à-vis the culture of the West as it was approaching the end of modernity.

Contents: Christian missionary thinking in its broad historical context · Explicitly missionary but non-Christian movements in Germany at the time (Hitler's missiology and Hauer's neopaganism) · Attempts in the US, in Britain and the wider ecumenical movement (William Hocking, Joe Oldham, the Oxford conference of 1937) at rethinking Christianity

Frankfurt/M · Berlin · Bern · Bruxelles · New York · Oxford · Wien
Distribution: Verlag Peter Lang AG
Jupiterstr. 15, CH-3000 Bern 15
Telefax (004131) 9402131

*incl. value added tax, the current german tax rate is applied
Prices are subject to change without notice
Homepage http://www.peterlang.de